DANVILLE PUBLIC LIBRARY
3 1205 00378 0697
P9-CSA-195

Heart of Dryness

Heart of Dryness

How the Last Bushmen Can Help Us Endure the Coming Age of Permanent Drought

JAMES G. WORKMAN

Walker & Company
New York

Published by Walker Publishing Company, Inc., New York

All papers used by Walker & Company are natural, recyclable products made from wood grown in well-managed forests. The manufacturing processes conform to the environmental regulations of the country of origin.

Botswana map: Sandy Gall
Map of water disputes in southern Africa: Peter Ashton

LIBRARY OF CONGRESS CATALOGING-IN-PUBLICATION DATA

Workman, James G.
Heart of dryness : how the last bushmen can help us endure the coming age of permanent drought / James G. Workman.
p. cm.
Includes bibliographical references and index.
ISBN 978-0-8027-1558-6
1. San (African people)—Social conditions. 2. San (African people)—Government relations. 3. San (African people)—Land tenure. 4. Indigenous peoples—Ecology—Kalahari Desert. 5. Droughts—Kalahari Desert. 6. Desert ecology—Kalahari Desert. 7. Water conservation—Kalahari Desert. 8. Kalahari Desert—Social conditions. 9. Kalahari Desert—Environmental conditions.
I. Title.
DT2458.S25W67 2009
305.896'1—dc22
2009005611

Visit Walker & Company's Web site at www.walkerbooks.com

First U.S. edition 2009

1 3 5 7 9 10 8 6 4 2

Designed by Rachel Reiss
Typeset by Westchester Book Group
Printed in the United States of America by Quebecor World Fairfield

To my father,
who provided sturdy roots,
and my mother,
who gave me restless wings.

It is scarcity and plenty that make the vulgar take things to be precious or worthless; they call a diamond very beautiful because it is like pure water, and then would not exchange one for ten barrels of water. Those who so greatly exalt incorruptibility, inalterability, etc. are reduced to talking this way, I believe, by their great desire to go on living, and by the terror they have of death. They do not reflect that if men were immortal, they themselves would never have come into the world.

—GALILEO, *DIALOGUES*

Contents

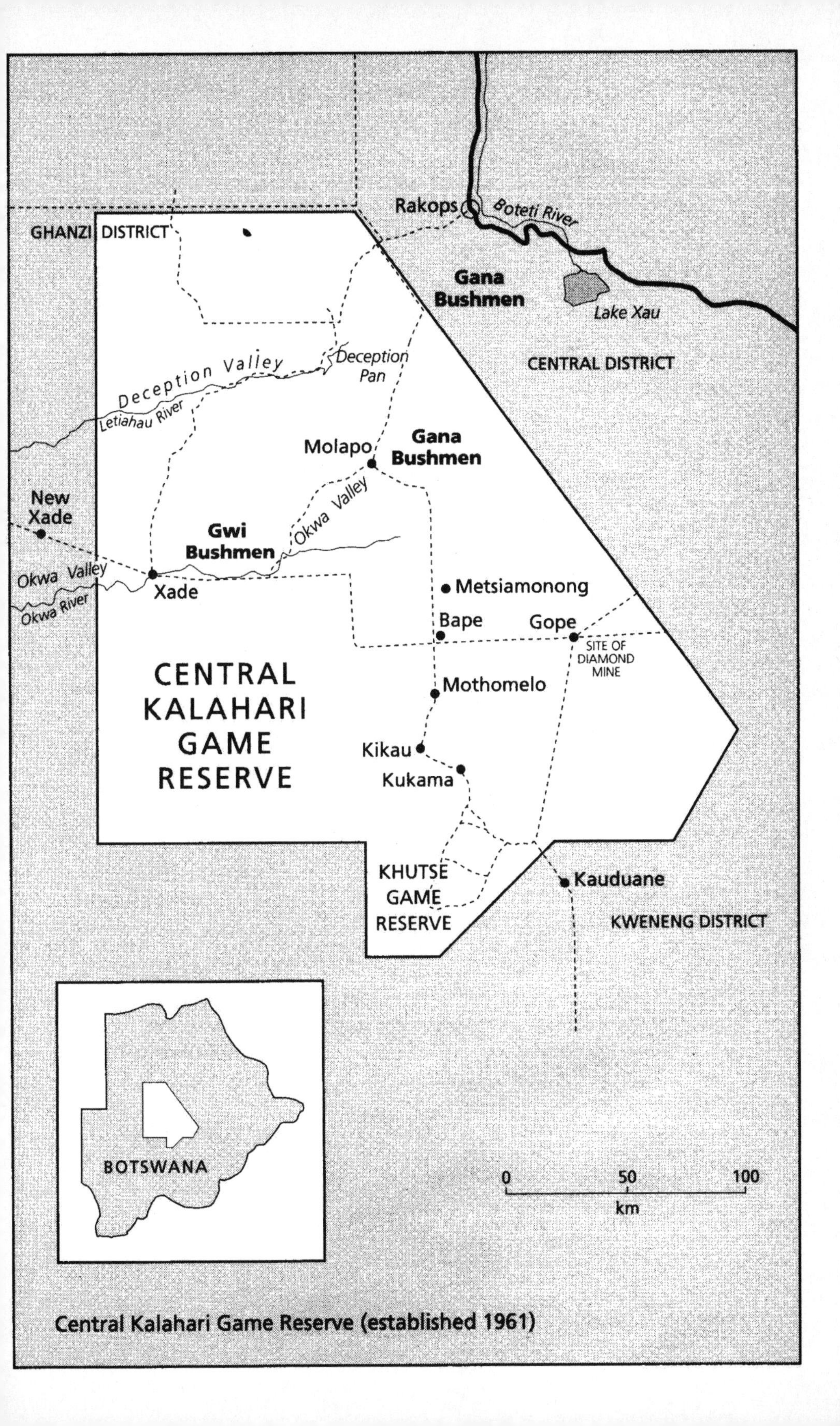

Central Kalahari Game Reserve (established 1961)

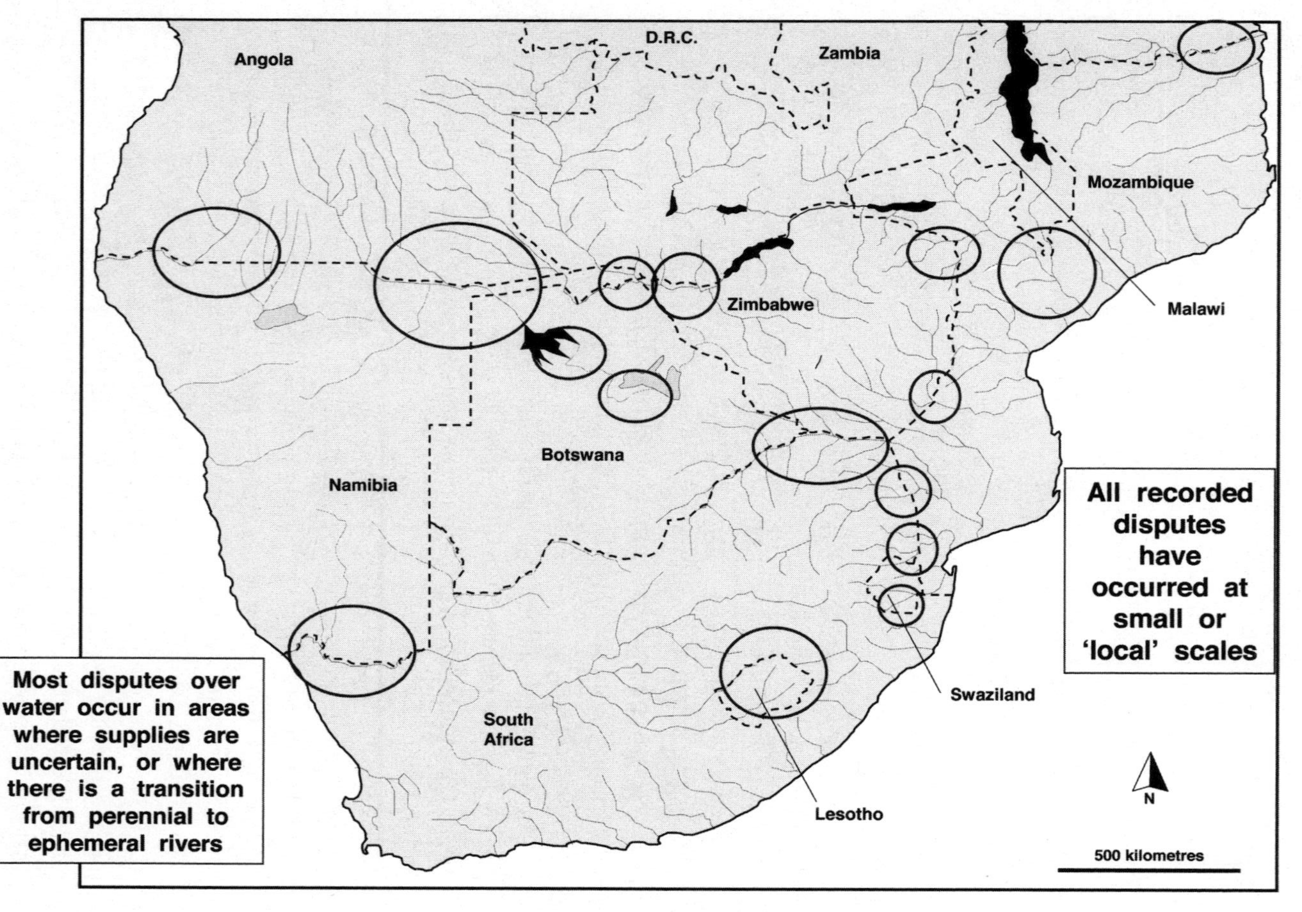

Water Disputes in Southern Africa

INTRODUCTION

To the Heart of the Matter

ONE STINKING HOT DAY DURING THE austral summer of 2002, the sovereign Republic of Botswana dispatched twenty-nine heavy trucks and seven smaller vehicles to converge on southern Africa's arid core. To reach their designated target, the drivers had to traverse one of the most kidney-jarring, axle-snapping, sand-blasted, and sun-burned landscapes on Earth. The destination lay at the heart of what local languages translate as "the Always Dry." Others call it "the Great Thirstland." On maps it is labeled the Kalahari Desert.

The convoy ground through flat savanna as drab bunchgrass and thorn trees rolled past the windows. Only the rare sight of springbok or ostrich broke the monotony. Eventually the vehicles crossed an invisible threshold and entered a territorial reserve inhabited by bands of indigenous people known as the Gana and Gwi Bushmen.

For tens of thousands of years Bushmen and their ancestors had thrived in this unforgiving landscape. According to geneticists, linguists, and ecological scientists, these people constituted the remnants of the world's oldest and most successful civilization.[1] But over recent centuries almost all were violently uprooted and displaced by better armed settlers: white farmers and ranchers encroaching from the south, black Bantu herders moving in from the north. Where half a million Bushmen once proudly strode the subcontinent as its sole inhabitants, barely a fifth of that number now lingered as an abject underclass. Many had intermarried or assimilated into the margins of the region's cattle posts or economies of Windhoek, Gaborone, Bulawayo, Johannesburg, and Cape Town. Scattered from their Kalahari homeland, Bushmen were typically relegated to a serflike existence, exchanging humility for charity. If the world largely wrote off Africa as a hopeless case, and if urban Africans dismissed rural tribes as ignorant and crude, even the poorest African looked down on chronically "destitute and miserable" Bushmen.[2]

For hour upon hour, the top-heavy vehicles jerked and careened forward as their fat wheels churned through the sand. That sand could reach 162 degrees on the surface, and in the peak of the day the heat expanded air pockets between the coarse grains, making the sand so soft and loose that even 4×4s bogged down. Drivers who let enough air out to increase rubber-to-sand traction increased their risk of a punctured tire. The maddening route grated on nerves already exposed by the unpleasant task they had to perform. And yet "it was not all gloom and sadness." Indeed, "the camaraderie lifted their spirits and brought playfulness to their character." Since "no one wanted to be a failure," the convoy made a game out of their assignment, and it became a "marvel to watch them display their prowess in attempting to outdo each other" as the jocular drivers raced each other toward the center of the Kalahari, unable or unwilling to turn back.[3]

Their assignment had been carefully mapped out in advance. Execution of orders was intended to be swift and unemotional. Sources would later differ about the degree of intimidation or physical violence involved, but some officers carried loaded weapons, for there could be no further negotiation with any of the remaining inhabitants.

Those still-intact bands of a thousand or so Gwi and Gana living at the center of the Kalahari felt confident in the ancient desert home from which they, unlike so many Bushmen, had never been driven. But even they owed their political asylum to international mercy. In the decade after the Second World War, upon learning how Israel was founded as a refuge for European Jews, Bushmen sought from England an equivalent for Africa's genocidal victims. "Listen to the weeping of a race which is very tired of running away," they pleaded. "Give us a piece of land, too. Give us a piece of land where our women will not be taken from us."[4]

In 1961 several British colonial officials leaned over a crude map of Botswana's sand-filled heartland, scratched straight lines into a twenty-thousand-square-mile trapezoid, and proclaimed the Central Kalahari Game Reserve. In doing so they drew upon the precedent of America's protected parks, refuges, and forests—wilderness areas set off-limits from development and kept uninhabited by people. But on this particular landscape, rather than seek a pristine virgin ecosystem, British officials planned to "reserve sufficient land for traditional land use by hunter-gatherer communities of the Central Kgalagadi" where the last surviving bands could develop on their own terms, free from relentless persecution to near extinction. Inside that hunter-gatherer haven, a thousand Bushmen clung fast to their autonomy in places with names translated as "Vulture Water," "Fossil Creek," "Kneel to Drink," and "Nowhere." Within their sanctuary Bushmen maintained a cultural identity unto themselves; they enjoyed a proud political autonomy that for various reasons annoyed and even threatened the more powerful surrounding Botswana republic. Outsiders referred to these central Kalahari Bushmen as the Last of the First—a distinction the approaching convoy planned to end.

Despite early popular reports of their existence in a "Lost World," Kalahari Bushmen were never mythical Children of Nature, sealed off from surrounding economies in an airtight bubble. Men occasionally walked out to work distant mines or ranches, while others trickled back and forth to exchange meat and skins for tea, tobacco, marijuana, blankets, and colored beads. Far-ranging hunters beat fence wire into arrow tips, pounded metal into spear heads, and made quivers from scavenged plastic PVC pipes. Like all of us, they adapted to available resources. Yet as the reserve's sole human occupants, Bushmen had not been overwhelmed by the currents of the outside world. If anything, small groups of semi-nomadic pastoral tribes like the BaKgalagadi, arriving four hundred years ago, had been transformed to adopt the dryland survival skills and indigenous intelligence of the Kalahari's original inhabitants. So the British mandate never intended "to preserve the Bushmen of the Reserve as museum curiosities and pristine primitives, but to allow them the right of choice of the life they wish to follow."[5] The enclave was to be their eternal refuge. If they so chose, proclaimed Botswana's founding father, Sir Seretse Khama, Bushmen could remain on their ancestral homeland "forever."[6]

Forever ended January 31, 2002. On that day Botswana's new president, Festus Mogae, unleashed a "hive of activity with Toyota Land Cruisers, Land Rovers and five-ton trucks" into the Kalahari's epicenter to sever the Bushmen's relative isolation and to merge them—one way or another—into the modern nation-state to which he decreed they must belong.

The final push toward forced assimilation had been building for quite some time.[7] Over several decades, tensions had been simmering between the dominant Tswana rulers and the Bushmen, an uneasy animosity more condescending than inter-tribal tensions with Herero, Matabele, Shona, or Kalanga, and more racially acrimonious than between whites and blacks. Some Tswana still considered Bushmen to be a distinct African species or regarded them as *malata*, meaning serfs, or even pets, livestock, and indentured servants.[8] Lately, the "modern" Bantu rulers' attitudes toward what they called "primitive" Bushmen had worsened, sinking from neglect to embarrassment to outright hostility until there no longer seemed to be enough room for the last ancient bands to remain in a landscape larger than Switzerland. That's when carrots extended to persuade Bushmen to move were replaced by sticks to coerce them. "How can you have a Stone Age creature continue to exist in the time of computers?" Mogae had demanded. "If the Bushmen want to survive, they must change, otherwise, like the dodo they will perish."[9]

The dodo did not perish in passive isolation. Its demise came as a direct outcome of the aggressive invasion by hungry and thirsty outsiders who displaced the indigenous species from its prehistoric habitat and dominated control of finite natural resources.[10] With the invasion of his convoy President Mogae determined to follow that precedent by targeting the most essential and finite natural resource, water.

Mothomelo, a settlement that lay at the heart of the Kalahari reserve, held the Bushmen's only borehole. For more than a decade a pump had lifted water from deep beneath the sand to provide a thousand people with supplementary water, distributed by government tankers to other scattered camps radiating out in all directions, where the water was stored until the next delivery. Upon arriving, the convoy executed its primary task. Trucks had unloaded their heavy equipment while officials tried to ignore the noisy protests of those odd people out here who in slang were "dwelling in the deep beyond the deep."[11]

Technicians sliced through the metal and dismantled the secure pump. A square steel plate was welded over the pipe until it was sealed permanently shut. Next, the officials overturned the plastic reservoir, allowing water to flow out until it was empty, and finally proceeded to crush it so that it could never be used again. Bushmen watched officials pour out their last, precious drops of water into the sand.

Eventually the convoy drove on through the sands to each of the remaining Bushmen settlements, where the government proceeded to complete the termination that it had undertaken and destroy the official supply of water it would never restore.

CUTTING OFF ANYONE'S water supply provokes a visceral response, and from outside the Kalahari came a furious backlash. Human rights groups launched a series of aggressive protests, vigils, boycotts, petitions, and lawsuits to save the last of those charismatic people best known from *National Geographic* essays or the film *The Gods Must Be Crazy*. Condemnation poured in from New York writers like Gloria Steinem, Hollywood celebrities like Brad Pitt and Angelina Jolie, supermodels like Iman and Lily Cole, rock stars like Jackson Browne, and, with less fanfare, diplomats from Europe, the United Nations, and the U.S. State Department.[12] Botswana's critics alleged the torture and genocide of Bushmen by officials who were motivated by lucrative ecotourism or blood diamonds, but against this noisy propaganda war the nation-state only dug its sovereign heels in deeper.[13] Driven by powerful incentives that it tried in vain to conceal, President Mogae's government proceeded to test the blameless Bushmen's endurance and faith in a series of losses seemingly torn from the Book of Job, and systematically stripped away everything that Bushmen valued most in the world. Officials hauled off their spouses, siblings, parents, sons, and daughters; then the government sealed off contact, halted medical access, burned their foraging fields, and spread cash, alcohol, and a mysterious deadly plague among idle, bored, and broken families. With the help of snitches paid to inform on old friends, it beat and crippled young fathers for seeking protein, fired shots into women and children, and ultimately imprisoned families within a tight perimeter, threatening death to those who crossed an invisible line. But perhaps this comparison is not entirely appropriate; throughout his biblical ordeals, even Job had been spared his water.

Naturally I sided with the Bushmen forced into oblivion, abandoned, left in the desert without government water. Yet among the continent's menu of atrocities even a bleeding heart must choose its battles. You cannot live long in Africa and be shocked—shocked!—to discover that well-connected elites of a dominant tribe here were exploiting weak and marginalized ethnic minorities of another.

The shock was to discover that Bushmen hung on.

A core group of young and old diehards showed no signs of backing down or surrender. Despite constant government pressure and water deprivation, a number of Bushmen clung fast to their sands. Even a rudimentary census is impossible, since many people would melt into the desert at the sound of approaching vehicles, but the government and Bushmen claimed the number of intransigent residents was as low as seventeen or as high as two hundred, respectively. They refused to be evicted. Through sheer cunning and savvy these stubborn Kalahari dissidents not only resisted Botswana's siege but did so during one of the hottest and longest droughts in the region's history.[14]

Over time I began to absorb the larger context and deeper meaning of their defiance. During its siege of the Bushmen at its waterless core, Botswana maintained superior force, technology, global communications, and financial resources. Yet global warming soon evened the odds. Even as President Mogae cut off the Bushmen's water supply, his own dams quickly began to vanish under nonstop scorching sun and rising winds that blew in hot gusts powerful enough to rip windsocks off airstrips. Each day, crucial federal reservoirs were evaporating above while pipes kept leaking millions of gallons below. With no streams to call its own, Botswana groped beyond its borders for rivers shared with equally parched neighboring states until border disputes erupted, armies mobilized, and the government found itself in the same disquieting position it had forced the Bushmen: landlocked, surrounded, isolated, desperate, and dry.

The Bushmen's quiet tenacity humiliated Botswana, but it also humbled me. Up to then my career had been spent trying to reform and modernize various nations' governmental control of water, and I had focused my efforts at the summits. I had exposed state water follies as an investigative journalist, crossed over to become an aide to America's top federal water bureaucrat, joined an unprecedented consortium to shape global water governance, and from there advised state officials from Argentina to India to Indonesia and China, collaborating with leaders from the World Bank and the World Conservation Union while helping Nelson Mandela chart a new global course for water policy. It was all rather heady stuff. Then along came these Bushmen to shatter my prejudices. By managing to cope without government water while drought crippled the surrounding state, the dissident Bushmen revealed the inherent fallacy of centralized water control. In the process they revealed my ambition—international water expert—to be an oxymoron. In the face of scarcity all water, like all politics, becomes emphatically local.

A few radical hydrologists were reaching similar conclusions, at least in theory. To eliminate water scarcity, went their hypotheses, government must force water underground, distribute water outward, decouple rate structures that suppressed water's real price, decentralize water use decisions, and devolve water allocation to the lowest level. This modest proposal might cost $1 trillion per year,[15] and assumed leaders might willingly surrender their power to convert the political economy to a hydrodemocracy in which water managed, regulated, and ruled its people.

In reality no civilization anywhere had the discipline to impose or tolerate such exacting self-rule. The one exception was the besieged Bushmen. I went to find them, in order to unlock their enduring code of conduct before it too was lost to the wind.

My motivations were hardly altruistic. In 2002 the most conservative vulnerability scenarios for the United States showed its groundwater declining, its snowpack shrinking, its infrastructure leaking, its dams choking with sediment, and its reservoirs evaporating. The Bushmen's present augured Americans' immediate future: the end of dependence on easy access to government provision of abundant freshwater.[16]

Looking ahead, the prospect appeared even more bleak. Temperatures in America were projected to rise one degree Fahrenheit every decade while the World Health Organization estimated that deaths from heat exhaustion would double with each degree.[17] Even as American politicians loitered in collective denial, the United States was absorbing the first spasms of an aberrant climate. Flash floods paralyzed sewage treatment plants, while heat waves baked topsoil and mushrooming populations collided with never-ending dry spells. There appeared to be as many ways for Americans to run out of water as there were definitions of *drought*.[18]

Far from representing a universal absolute, drought typically describes a subjective and relative condition that varies markedly by population and place over time. A farmer's curse is a vacationer's blessing. An Arizonan's rainy year is an upstate New Yorker's dry spell. Even under constant annual weather patterns, runoff that once sustained twenty thousand inhabitants would be regarded now, in million-strong cities, as a crippling drought. Set against the background equilibrium patterns of previous decades or centuries, measurable drought occurs whenever mean temperatures escalate hotter, water tables sink deeper, evaporation rates accelerate faster, runoff shrinks lower, reservoirs vanish sooner, dry seasons last longer, or economic thirst of more people demands more water for more activity than ever before. Of course, nothing prevents all these unpleasant forces from compounding at once, and the consensus of scientists suggests that in the United States we are now entering precisely that convergence.[19]

This scenario is known as a perfect, perpetual, or permanent drought. It means that in spite of unprecedented prosperity and freedom in most other sectors, and in spite of the undeniable convenience of cheap running taps and $8-billion worth of bottled water on supermarket shelves, Americans enjoy less absolute access to water than ever before.

Consider demand. When Europeans first arrived in North America, fewer people consumed less water. The New World held at most 14 million noncentralized people, each annually consuming perhaps 100 gallons each year. Today each of 300 million Americans consumes on average a daily 150 gallons, plus another 5,000 gallons indirectly through food or industrial goods.[20] Human demand rises with affluence as we move to bigger houses with pools and gardens and eat our way higher up the water-intensive food chain. Most of us pack into cities highly dependent on deteriorating water systems that were designed for significantly lower needs in historically wetter eras.

Against rising thirst, supply has been correspondingly decimated. Due to pollution and population growth, each American can "own" less than half the freshwater of the previous generation. Collectively, *Homo sapiens* has taken so much from all other species that aquatic biodiversity suffers the highest rate of extinctions. Yet the average U.S. citizen still competes for smaller quantities of freshwater over time as our individual shares have shrunk.[21]

Worse, many formerly cool wet places are rapidly warming up or drying out. Fryeburg, Maine, gasps as a foreign bottled water conglomerate removes water in ninety-two tankers per day,[22] while New Yorkers clash with upstate farmers over tributaries that allow Gotham's existence, and some Appalachian residents can't drink from taps.[23] In Minnesota, Iowa, and Nebraska dozens of parched towns that sought water are blocked from extending pipes into the thin Missouri River, and deep wells sunk from South Dakota to Oklahoma have tapped out most of the Ogallala Aquifer.[24] Abnormally arid conditions have dried up the Pacific-bound Colorado River and Atlantic-bound Rio Grande before either can reach the sea; have desiccated all or parts of forty-nine states;[25] and have shrunk fifty-seven rivers beneath record lows.[26] For relief, dry frontier populations once dispersed to wetter landscapes, but within a few years thirty-six dry states could have no place else to go.[27]

Our predecessors stored water and stockpiled food against hard times while waiting for cyclical dry spells to break. That strategy may not work much longer. Paleoclimatologists discovered that the prosperous epoch since 1615 was unusually humid; the intervening four centuries, especially the rainy twentieth, brought the wet exception to North America's chronically dry rule. It turns out that urban water allocation, rural water rights, and agricultural cropping patterns were all set during a few decades of abnormally heavy snowpack and rainfall. Now researchers project that dry spells could linger for decades or indefinitely. The first American colonies were crushed by a few successive arid years out of the last eight hundred. Today we are facing the worst hot dry era in thirty thousand years.[28]

In dry times the New World relied on global aid from and trade with damp Old World cities and farms. Today those formerly reliable overseas breadbaskets have grown more parched than our own. London now imposes water rationing, and the rain in Spain falls nowhere on the plain or anywhere else in that country. Among our trade partners, Mexico is literally collapsing on top of empty aquifers, while Canada's heartland faces unprecedented stress and China lacks enough water to drink or feed itself.[29]

As the reality sank in about rising thirst and falling water tables and aquatic extinction, the usual suspects—hand-wringing liberal environmentalists and social activists—cranked out their litany of doom, right on schedule. Only this time their worry was amplified by a chorus of nervous conservatives, industrial interests, and development boosters all agreeing that the end of water was nigh. Never mind the gloomy Club of Rome;

bullish corporate titans at the World Economic Forum in Davos warned us how limits to growth came from scarce freshwater.[30] Elbowing past picket lines of wild-eyed green protesters, some otherwise astute World Bankers were confounded at how, contrary to basic economics, water use grew less efficient as water grew scarcer.[31] Goldman Sachs, Wall Street's sober investment bank not given to hyperbole, grew increasingly alarmed that, by doubling every twenty years, global water consumption had attained an "unsustainable" rate of growth. Two out of every five global industries believed that the impact of a water shortage on their business would be severe or catastrophic, and fewer than 17 percent said they were prepared for that calamity.[32] Even innovation-driven optimists worry how California's information technology firms require 3,400 gallons per microchip, a quarter of Silicon Valley's dwindling reservoirs. The high-tech panacea, desalination, is so energy-intensive that it relies on power plants that now require 40 percent of existing freshwater supplies.[33]

Yet the real threat to the United States from water scarcity may be less economic than political. Most uncontrollable external disasters—floods, hurricanes, wars, tsunamis, earthquakes, terrorist attacks—bring out the best in Americans, pull our can-do society together, and motivate us to roll up our sleeves in mass collaboration. Drought splits us apart. It pits you against your neighbor, and your state against mine. Our country's most experienced water managers—the men who ensure we have enough to drink—have confessed they lack a solution and, looking ahead, say our common future may resemble "an Armageddon."[34]

Following decades in which armies fought over oil, U.S. security analysts are proclaiming, "Water is the new oil." It is in fact substantially worse. Countless energy alternatives let us move "Beyond Petroleum," but water has no substitute. No dehydrated economy can exist. After three thirsty days you secured clean freshwater, or you died.

Or you killed.

In places like Bolivia, Kenya, and Iraq the strong who controlled access to water killed the weak who increasingly sought it.[35] Asia's drinking water kills tens of thousands from dissolved arsenic or heavy metal, while post–Green Revolution farmers kill themselves when their wells run dry.[36] Worldwide 1.8 million die annually from thirst, dehydration, and waterborne diseases, and during the last violent century the lack of clean water killed more people than all the casualties of all the wars combined.[37] Now our military prepares for the onset of wars fought over water itself.[38]

Conflict over water in the New World predates Columbus. Unfortunately, as acute water shortages worsen, the United States has deliberately severed the last safety net that avoided or delayed catastrophe in the past: drought-adapted indigenous people. Better-armed European settlers used to beg, borrow, and steal subsistence services and survival skills from the intricate network of Native Americans, whose innovation lay in social and natural coping mechanisms. Since then, extant tribes have been dispossessed by and absorbed into immigrant populations. Those who still hold fast to the traditional ways do so more from pride or nostalgia than from necessity. Even the few rare victories—a new casino license to win back from Europeans a fraction of the tribal wealth stolen over time—take the young further away from the ways of their ancestors, and a similar fate has befallen the drought-adapted aboriginal people of Australia and Africa. Almost all dryland indigenous bands have been co-opted into industrial economies: wiped out by disease, forced on government-run reservations, hooked on the irresistible lure of food in cans and alcohol in bottles and water on tap.

For more than a decade even the wildest drylands of Africa no longer held autonomous bands who might share their self-sufficient experience. Then Botswana's convoy destroyed the last government water supplies and deliveries inside the reserve, triggering their crisis—and my opportunity. I saw America's fate inextricably linked to the predicament of a thousand indigenous people suddenly forced to submit, die, or adapt once again to the Great Thirstland. The survivors had to tap into the deep reservoir of indigenous wisdom, and I hoped to grasp the essence of their unwritten code. For centuries Bushmen had been shot and infected, poked and prodded, and now, as they faced the onset of permanent droughts, I set out to exploit the survivors one last time.

The "Last of the First" welcomed me to their fire. I listened to what often seemed serious debate but was later translated as spectacularly lewd banter. During a lull one evening, as it grew cooler, I moved with tape recorder and camera from one Bushman to the next until coming to an unspoken matriarch. In exchange for smuggling contraband water and other supplies, I sought to extract from her and others a few Important Answers to Big Questions, namely, "What will you do without government-supplied water?"

She kept scooping flesh out of a tsama melon, trading gossip with another.

"How are you going to manage water during the drought?"

The old woman shrugged without looking up and shifted back on her heels. Next to her a small fire burned. It was more smoke than flame but never seemed to go out.

I persisted. "Do you think you could manage enough water for your family and your band to last until the rainy season?"

Like others before her, she grew evasive. Repeating the question through my translator met with awkward silence. Back then, her caginess didn't make sense. Years later it began to. It wasn't that Bushmen didn't want to answer; they just couldn't. An "international water expert" grilling the old Bushmen about how humans must manage water was like a Vatican cleric interrogating Galileo about how the sun must orbit the earth.

To be sure, we will not soon abandon eBay or Wal-Mart to hunt and gather in foraging bands. Nor should we feel the need to. Yet the Bushmen code of conduct may help us escape a Hobbesian or neo-Malthusian nightmare. Prepared for extreme deprivation, Kalahari Bushmen chose the hard responsibility of a dry reality over a government-dependent fantasy of water abundance. Outside of their reserve the so-called civilized world found that for all our military might and Internet bandwidth, certain things still lie beyond our grasp. We discover we cannot "regulate" barren rivers and depleted aquifers any more than we can "regulate" our climate, clouds, or rain. Out here, while elected leaders kneel to pray for a thundershower that will provide temporary relief, the increasingly dry hot wind whistles through the thorn trees in the central Kalahari and whispers the ancient secret those last defiant Bushmen never forgot.

We don't govern water.

Water governs us.

PART I

Collision

CHAPTER 1

Kalahari Rivals

A MAGNIFICENTLY WRINKLED FEMALE Bushman squatted by a small fire, stoking the embers beneath a smoke-blackened kettle. To the outsider, the creases of her light reddish skin and the folds of her faded and filthy scraps of cast-off clothing carried the mixed aroma of dust, woodsmoke, and stale sweat. No one else seemed to notice the sour odor—the tight-knit group had long grown accustomed to the smells of one another.

The woman was in her sixties or perhaps seventies. No one knew her exact age, not even she. Elder Bushmen marked time by stories linked to the shape and position of the moon. They didn't honor birthdays. In any case she was not youthful or what most consider pretty. She had low cheekbones, a bony chest with empty breasts, and lacked curves in her waist, her hips, or her buttocks. Deeply creased lines furrowed across her brow and drooped from her Mongoloid eyes down to her jaw, avoiding the wide space of skin stretched between her nose and upper lip. She appeared to have no brows above deep-set eyes that took in everything. Her hair was short, dirty, and turning slightly gray. Both her shirt and dress were torn open. Her square features looked almost masculine. Even with her black and white bracelets, and hoop earrings, a glance at her photo led foreigners to think she was a man. But in the way she moved and by the evidence of the offspring to whom she had given birth, she was entirely female.

Her first name could be spelled out *Qoroxloo*.[1] To say it out loud, lightly inhale the first consonant click, drawing the tongue down and back from the roof of your mouth. Swallow the *r*. Exhale and pucker the second *xl* click, moving the tongue in from the side of your jaw, laterally, like the sound to move a horse. And now: *Duxee*. The surname's *x* requires a dental click, pulling the tongue tip back from top front teeth. It's a tongue twister, and I invariably mangled it.[2] Regardless, one should try to pronounce her name out loud, like a soft prayer, to affirm her ancestry and prevent her existence from growing as homogenized and anonymous and assimilated as her life refused to become. Though she had never mocked my attempts to speak in clicks. Her face had only wrinkled up even further as her eyes suggested mirth.

Perhaps some deficiency in outer beauty made Qoroxloo Duxee's heart overcompensate. She had worked hard as a young girl, learning quickly from the elders. After listening carefully to the stories and lessons of her ancestors, she kept her hands busy. Few ever heard her complain. She bonded easily, with thoughtful gestures and generous gifts, and shared everything new she discovered. Because lingering grudges could prove fatal, she embraced physically and frequently; she forgave quickly and permanently. Mohame Belesa, her husband from an early age, came to love her with loyalty and affection. Decades later, in midday shade or at night by the fire, the couple could usually be found sitting side by side in the sand, an arm or a leg overlapping the other in that easy relaxed fit which grows from a lifetime spent together.

Despite her age, and the trials she endured, Qoroxloo remained exceptionally energetic. She was constantly teasing, aiming her sharpest barbs at those closest to her. Many in the extended band looked up to her. Younger women and children sought to emulate her behavior, and even men might listen thoughtfully to what she had to say. Yet Qoroxloo never told people what to do; she made suggestions that she would proceed to follow herself, allowing others to draw what they would from her example. Unlike tribal chiefdoms or urban bureaucracies, no title came with her unspoken stature.

In recent years, Qoroxloo noticed gradual changes in the young. Some had picked up some English and Setswana, new languages that allowed interaction with foreigners, but that also turned them outward, away from the old ways. Sent to schools, they enjoyed comfortable beds and tap water, and when they returned a few carried shiny tools which, they explained, allowed them to speak back and forth to unseen people.

To the ancestors? You speak to them?

No, Grandma, living people. Here in the heart of the Kalahari she only heard scratchy noise, but they told her how in "less remote places" the tools "let you talk with anyone on earth."

Qoroxloo did not consider her landscape remote. To her it was not excessively "harsh" or especially "dry," and she could not understand why anyone could be afraid or uncomfortable in it. It was the only habitat she knew, though she was aware some larger world extended beyond the edge of the horizon, farther than even she had walked. Foreign travelers described for her oceans, green landscapes of steep mountains and raging rivers with something called fish. But she could not fully picture them, even when shown photographs. She never expressed a desire to seek them, for everything she needed lay around her, until one day, at the start of the long, dry season, which brought something she most certainly didn't need.

Her band fell silent and listened as the air carried the sound of heavy engines grinding toward them through the thick sand. Trucks, they decided, including a few big ones. Qoroxloo had lived her first several decades without ever hearing a motor. Later, vehicles remained rare, but mostly welcome diversions, bringing unusual stories and white and black people from the outside world. Most were researchers or officials, and some brought useful gifts, like knives, scissors, or wire. But in recent years the sound of certain engines had grown familiar and menacing; it was a sound that brought relentless pressure from men in uniforms, men with guns. Men telling Qoroxloo's band that they could not live here any longer, that this was no longer their proper home.

As the government vehicles approached, driving faster than necessary, Qoroxloo told her family that this could be the final termination of services. The government at last was bringing the end of its water and the start of something else.

A few officials later testified in court, under oath, how they merely had set up tables on the periphery of each settlement, then patiently waited for Bushmen to volunteer to depart. Never had anyone used force or intimidation. But hundreds of eyewitnesses, and film from a video camera that surfaced later, recorded a picture that differed, as a judge put it, "as day from night."[3] Visual flashes of the ensuing chaos showed the destruction and dismantling of huts and water supplies.

Qoroxloo and her band watched as officials entered their compound, broke open ostrich eggshell canteens, dumped out plastic storage containers, and overturned steel barrels that all contained water. As the government's agents poured families' water into the hot, deep, dry sands, Bushmen saw their lifeblood spilling out. Children yowled and clung to screeching mothers as men watched in impotent, silent fury. The armed officials said little but duly carried out instructions until it appeared there was no water left anywhere.

The methodical cutoffs were of course only a means to an end. As the spilled water began to evaporate, truck drivers offered rides out of the reserve. They were willing to carry away any thirsty people who were now, suddenly, ready to leave behind forever the only home they had known. Free of charge, the government prepared to haul off families, huts, and worldly possessions to live the good life in one of three designated resettlement areas named Kaudwane, Xere, and New Xade. Many officials said they were doing Bushmen a favor, and sounded genuinely earnest in their sales pitch. They assured Bushmen that life was easier outside the reserve; life was better in resettlement areas; life was new, comfortable, improved, and mainstream, all awaiting outside once you "voluntarily" uprooted. Collective negotiation was forbidden, however, and officials separated and singled off brothers from sisters, husbands from wives, children from parents. Just sign the paper or, as few Bushmen could write, a thumbprint sufficed. Press there, on that dotted line, and the government would provide free schools for children, free birth control, free clinics for the sick, free cows, cash compensation, and, when the time came, free funeral services.

Above all, in exchange for shrugging off the old ways, the government offered Bushmen water—free and abundant access to an endless flow of life: as much convenient wet stuff as Bushmen wanted for as long as they ever wanted it.

For years Bushmen had declined such seemingly benign offers. They had heard what happened to others who had been relocated outside the reserve years before. Word trickled back through the bush telegraph of sickness, boredom, whoring, and violence. People who were used to ample space, to roam freely and self-sufficiently, compared the resettlement camps to "being cooked in a cooking pot." Others called New Xade "a place of death." But once the government shut off delivery and destroyed their water supply, they said, the old unwelcome message sounded different. It became an offer few could refuse as the dry season loomed ahead.

On that day many Bushmen decided they should cut their losses and just go now. Take the cash, the cow, the free water. Instructed to gather up their belongings and utensils, Bushmen found admittedly little to take: some clothes, bags, blankets, a few digging sticks, snares, bows and arrows, spears, and some containers. Most wondered what they could use these tools for, since where they were going allowed no hunting or gathering. The ample water promised to them out there left no need for water containers, springhare bladders, plastic bottles, or ostrich eggshell canteens.

Qoroxloo's husband returned to her side during the commotion, but there was little to say. Qoroxloo saw the young, usually so excited about new things and so full of their sense of immortality, begin to waver. She understood the pressure on them to depart, especially for those who had lost sight of the ways of the ancestors, those who had grown far too accustomed to official water deliveries, even addicted to them. In one of the few times she revealed her anger Qoroxloo lashed out with her belief that the government had initially provided Bushmen with water to break their resistance and will. "When I was young," she said, "the men hunted and we got our water from the roots of plants. We lived well, and people only died of old age, not of diseases. But then government officials started bringing us water and mealie-meal; they kept bringing these things. And now that I think about it, I believe it was to make us dependent, and abandon our traditional ways, and get us to move."[4]

The trucks revved their engines. Qoroxloo saw other Bushmen turning, looking at her. They wondered what she would do, whether she too would give in and depart. She had amassed a lifetime enduring water scarcity, but this confrontation was unprecedented. Larger political pressures were prodding her band in their own sanctuary. As the officials moved from family to family, loading up people she would never see again, Qoroxloo did not join them. Seeing no real alternative, she chose to remain. At the same time, she failed to conceal outward signs that she, too, was anxious and more than just a little bit afraid.

AS FAR AS I could tell, the Bushmen have no indigenous translation for *rival*. The word is apparently a Western construct, born in ancient Rome. Intriguingly, it shares with *river* the same Latin root, *rivalis*, or: "one who shares the same stream or water."[5] Like drought, rivalry seems to emerge from a relative condition of scarcity and ecological stress, whether the source is a well, a creek, an aquifer, or a spring. In rainy places neighbors had leisure to calmly decide how they would access, allocate, and share water with each other from a position of comparative abundance. In arid landscapes, however, as rainfall declined and runoff diminished, latent rivalry would invariably rise to the surface and explode. Thus in a chronically dry country everyone is a potential rival, and the extent of their rivalry increases in proportion to the proximity and threat to your water. In a democratic republic like Botswana, where government owns and controls all resources and decides whether to share that water or one day cut it off forever, there were many men and women just doing their job, following orders. But at the top of this chain of command, Qoroxloo really had only one accountable individual who became her rival.

Though not a tall man, President Festus Gontebanye Mogae cut an elegant and well-manicured figure. His heavy-set frame favored pin-striped tailored suits, with cuff links, and a silver wristwatch. Outside his windows rose gleaming modern glass office towers, showcasing Gaborone, considered the fastest-growing city in the fastest-growing economy in Africa.[6]

Mogae had immense plans. He believed Botswana was on the threshold of great things. His well-governed nation, the oldest democracy on the continent, held unique assurance for the world, whose most powerful leaders shared his vision.[7] Bill Gates spoke of Botswana as a "model nation." During a March 1998 visit, Bill Clinton proclaimed "the promise of a new Africa whose roots are deep here in your soil, for you have been an inspiration to all who cherish freedom." Standing before five thousand gathered at soon-to-be-president Mogae's statehouse, America's forty-second president concluded, "Africa needs more Botswanas, and America is determined to support all those who would follow your lead."[8]

Mogae took immense pride in Botswana's lead and held his country to a higher standard. Far from a lawless state or socialist banana republic, Botswana was a more peaceable capitalist country than many in the so-called developed West. True, its population suggested a city rather than a country. But 1.8 million people had elected their foreign-educated and cautious leaders to govern a landmass larger than France through careful, gradual consensus. The contrast with Mogae's postcolonial African rivals could hardly be more glaring: Belgians had left Congo with 90,000 miles of good roads, which degenerated over three decades down to 6,000; the British had left 8 miles of paved roads, which Botswana expanded to 6,000. During the past forty years since its serene birth, Mogae's sovereign state had brought per capita income to $5,900—higher than Malaysia, and yielding more positive stability than in, say, Portugal or even Mississippi.[9]

As president, Mogae planned further to diversify a market-embracing economy that for decades had led the world in per capita growth. His nation's accounting books were tidy, transparent, and in the black; inflation was moderate, taxation low; and he had insisted on avoiding debt like that which hamstrung America.[10] Botswana's educated government workforce was happy. His Botswana Democratic Party ruled unchallenged. Botswana's pious and proud urban population—of *The No. 1 Ladies' Detective Agency** fame"—loved to shop; the capital city's dozen or so busy malls and multiplex movie theaters could be exchanged with any from Gainesville, Florida, to Bend, Oregon. To build its nonracial nation, Botswana was determined to erase tribal identity and ethnic distinctions; its black founding father, Seretse Khama, had transcended prejudice to marry a white foreign wife. No other country, not even the United States, had risen so far, so fast, from so little.[11]

For good reason Botswana was frequently hailed in international magazines, feted at international economic forums, and favored by international tourists. Top statesmen at New York's prestigious Council on Foreign Relations praised President Mogae as a "visionary" who calibrated everything just right, like a chess master.[12] In early 2002, Botswana's sovereignty was absolute, its reputation pristine.

Mogae's disciplined military had never fired a shot in anger at foreign or domestic enemies throughout its existence. It deserved its reputation for pacifism. Yet it was not always necessary for a president to ensure national security or promote domestic tranquillity through the barrel of a loaded gun. The desert landscape offered a more subtle weapon that might suffice.

Indeed, on the very day President Mogae hosted a large UN conference

---

* Alexander McCall Smith is the prolific writer of this internationally bestselling detective series, which portrays the bourgeois character of the nation's citizens in a kind, if somewhat dated, light. In his novels McCall Smith has his Tswana protagonists adopt two Bushmen children abandoned in the Kalahari; in real life the author has spoken out against the government's policies.

on "peaceful and constructive group accommodation in situations involving minorities and indigenous peoples," he was destroying the Bushmen water supplies to expel them from their Kalahari.[13]

His water cutoffs initially escaped irony, or wider notice, for in early 2002 the world's attention was transfixed by upheaval in Afghanistan and, soon, in Iraq. In comparison to more lethal global injustices—war, disease, riots, famine, pogroms, genocide, and terrorism—the concurrent fate of a thousand illiterate African foragers was geopolitically insignificant. President Mogae portrayed his exercise in the Kalahari as a domestic development scheme, the kind of replumbing operation that in Asia forced the relocation of thousands each day. Officials were openly baffled by any question of impropriety. Botswana had elected Mogae to serve the greater common good. His party wisely had invested surplus wealth in health, education, roads, and vital public water infrastructure—including the borehole, pump, and storage tank that it had, that very day, obliterated.

Mogae's policy of water curtailment had ample precedents. Ruling governments dating back to ancient Rome and imperial China had established this tactic that was later (as will be shown) practiced by nations as diverse as Bolivia, Israel, Iraq, Germany, South Africa, Sudan, and the United States. Throughout time—and invariably in the name of state security—government deployed water as a weapon to coerce insubordinate people.

The logic is simple. Humans can live only a few days without water. Control water, and you control every aspect of those of us who depend on it: how we live, where we live, and whether we live.

The Kalahari offered an ideal stage for Mogae to apply such a tactic. The landscape was hot and dry and held no reliable standing water—no rivers, no lakes, no springs. So government officials concluded, reasonably enough, that the Bushmen would ultimately give up their primitive ways, accept the state's generosity, adopt the modern economy offered by Botswana's civilized culture, and at last submit to the benign will of a higher enlightened authority. Mogae's friend, the managing director of Botswana's diamond company, wholeheartedly supported the eviction of Bushmen from the Kalahari reserve for their own good, "Otherwise who would always want to remain in the Dark Ages while others move forward?"[14]

During what climate scientists later recorded as the planet's seven hottest years, the loss of water should have forced the last free Bushmen to surrender and assimilate. It should have brought an end to "the harmless people" and within days broken their will. With only dry fossil riverbeds, Botswana's government officials expressed their confidence that any stubborn remaining dissidents would soon trickle out of the reserve of their own accord in the arid days ahead, or perish like their president's proverbial Dodo. Through his simple, decisive act President Mogae planned a quick end to Africa's complex and intractable centuries-old issue that had come to be known even officially as the "Bushmen problem."[15] He planned his water cutoffs to be a final resolution to a long and messy debate, not caring whether those Stone Age people might have categorically different plans.

CHAPTER 2

Crossing the Threshold

IF THE HUMAN BODY IS CUT, PLATELETS gather to clot, creating a barrier against blood loss and infection to a flesh wound. If invaded, insect colonies repair their home's protective seal. But these natural systems require complex, coordinated signals. Against the government's Kalahari convoy no centralized nervous system or communications infrastructure existed among dispersed bands of Bushmen. Natural aridity of the desert itself had always been their main line of defense against potential aggressors. Now, as they faced the encroachment of powerful external threats, their intrinsic strengths became weaknesses.

Qoroxloo's people had evolved in autonomous face-to-face groups of thirty to fifty individuals, and even the larger bands seldom grew beyond 150. Each found its own equilibrium and order through small-scale, nonhierarchical self-regulation. Their human nature was of course the same as ours. But internal strife, jealousy, and tensions were relieved from within, avoiding lawyers, courts, or police enforcement. For millennia they had enjoyed the Kalahari core all to themselves without competition. But as Botswana split up their families and terminated services, Qoroxloo and others had to seek, for the first time, political representatives to speak and act on their behalf.

Roy Sesana, also known as Tobee', was the Kalahari's last surviving radical activist, a charismatic figure who along with six others had formed First People of the Kalahari. FPK was a Bushmen group ready to fight for indigenous rights to resources like water, and shrewd enough to oppose armed force through nonviolent resistance. Even if exiled by thirst, Sesana urged his people never to sign away their birthright. Instead he would seek other ways, and overseas allies who could help the Bushmen fight in court.

Botswana's politicians mocked his claims. They tried to dismiss his lawsuit on a technicality: Sesana could not represent thousands of Bushmen, since these bands lacked headmen or chiefs. They had no elections, no formal conflict resolution, no judges, hereditary figureheads, parliamentarians, or king. They epitomized autonomous and egalitarian decision making.[1] So who was Sesana to put himself forth? "I am a leader," he calmly replied, further stating that when he "was a boy we did not need leaders and we lived well. Now we need them because our land is being stolen and we must struggle to survive. It doesn't mean I tell people what to do, it's the other way around: they tell me what I have to do to help them."[2]

As Sesana visited Kalahari camps, Bushmen told him to fight for their rights.

He listened. In Kikao, Sesana listened to a smart and savvy young man caught in the no-man's-land between the old ways and the new. Jumanda Gakelebone grew up drinking cool water from an ostrich shell and trance-dancing naked around the fire, but also later learned how to work a cell phone, negotiate city streets, speak English, and resist the urge to drink alcohol. The energetic Jumanda teamed up with the battle-scarred veteran, Roy, who in turn forged alliances outside of his home band with Bushmen elders, lawyers outside of the Kalahari in Ghanzi or Gaborone, and with the activists in cities outside of Botswana who might be reached by phone or even, as Jumanda had recently discovered, through the screens of computers in one of the country's isolated Internet cafes.

In addition to tapping into old ways of survival, Bushmen learned a portfolio of new skills in a hurry. They developed and reinforced physical, legal, economic, and ethical practices all based on securing access to water. When the government described its water cutoffs as *voluntary resettlement* and *development*—words that could not be translated into their language—the Bushmen replied that it was a tool of forced eviction and submitted evidence to prove it.

Initially Bushmen tried the legal route, as 243 Bushmen signed a paper whereby human rights attorneys would represent them collectively in a class action suit against Botswana's government. Africa's regional appeals court ordered Botswana's high court to try the case in full and to give its verdict with "all due process." Both sides anticipated the case would be quickly dispatched: three weeks in Ghanzi to hear Bushmen plaintiffs; two weeks at the high court to hear the government defense; resolution in time for Bushmen to return home "before the next rainy season."[3] No one anticipated that a four year-and-ten-month megadrought would pass before any final judgment.

In the meantime, as Bushmen found a collective voice, they knew possession remained nine tenths of the law. So while Bushmen leaders fought externally for their day in high court, a core of dissidents who hunkered down inside the Kalahari reserve would seek victory in the global court of public opinion. If Bushmen like Qoroxloo could endure the pressures and isolation to vote with their intransigent feet, their examples would carry immeasurable political weight.

Roy Sesana wasn't the only Bushmen leader orchestrating resistance; his charisma and penchant for drama was balanced by Mathambo Ngakaeja, a soft-spoken Nharo Bushman who seemed to be constantly nudging spectacles up the bridge of his nose. As the first Bushman university graduate with a degree in geology, Ngakaeja had been wooed by the government and its mining companies with dazzling starting salaries. Instead, by leading Botswana's Working Group on Indigenous Minorities in Southern Africa (WIMSA), he often challenged those same would-be employers, albeit in a less confrontational way than Roy. This time even Ngakaeja was outraged and maintained Kalahari Bushmen could only live free if they could secure access to water. "Obviously if you have no water to drink and you have no water to give your children, you will be forced to go to the next place where water is."[4]

To secure that water inside the reserve, Sesana and Ngakaeja teamed up and prepared a resupply vehicle. Yet even a Bushmen movement held potential cracks. President Mogae's men had sponsored Bushmen tipsters to report any unusual dissent or suspicious behavior. The snitches later confessed they hated informing on their colleagues, who fell silent in their presence. Nevertheless, word leaked up the chain of command, the government intercepted the relief vehicle at the perimeter of the reserve, refused it entrance, and turned it back. An invisible wall had sealed off the heart of the Kalahari to any outsiders but high-paying tourists.

About that time I wandered into Botswana's capital city, seeking interviews with the Bushmen leaders and their legal representatives. In desperation they saw my Land Rover, read my sympathy, and asked me to pose as a tourist and smuggle reinforcements to Qoroxloo's band.

I hesitated, partly from journalistic detachment but mostly out of cowardice. The president's cabinet had threatened severe reprisals, resenting "undue foreign intervention," and had explicitly "rejected any and all outside assistance" for Bushmen. I feared the real and imaginary risks I might face. I was by nature an observer, and by crossing the line from note-taker to participant to join some futile resistance, I would not only lose a reporter's impartiality but could also be arrested, deported, or jailed and forgotten. For decades I'd avoided getting entangled in hundreds of other worthy causes, yet this ordeal seemed different; it had a "last stand" feel to it. In the course of my interviews with evicted Bushmen outside the reserve, more than one Bushman told me, through translators, "I only want to go back to my home in the Kalahari, live there until I die, and be buried in my ancestral land where my father was born and is buried."

The next morning Bushmen leaders transferred their illicit "contraband"—sacks of cornmeal, sugar, stew, dried fruit, tea, tobacco, medicine, and water—into my vehicle. We buried it all beneath two weeks of dirty laundry, the stench of which should discourage any inspector.

"You sure you're okay with this?" Ngakaeja asked.

"No one's forcing me."

"Have you been in the desert before?"

"No," I replied, masking anxiety under false bravado. "But hell, I've got a decent map. Where am I going?"

Ngakaeja spread out the map on the hood and instructed me to drop a third of the supplies in Gugamma and two thirds in Metsiamenong, farther to the north. We looked again at the map. It showed that there were eight settlements, including Gope, Kikao, Mothomelo, Bape, Xade, and Molapo.

"What about the people in these other settlements?" I asked.

Ngakaeja glanced up, nudged his glasses up the bridge of his nose, and then looked away. "There are no people left in those other settlements."

On the route north four paved lanes became two, then one and a half, and then nothing but that deep, soft, miserable sand. After hours of silence, I eventually rolled past a wooden sign indicating I was near the reserve's southern entrance at "Khutse," which in the language of the evicted Bushmen inhabitants meant "place where you can kneel to drink." If only it were still that easy. The state's hydrosocial contract had been irrevocably severed.[5] Never again would Botswana provide, or even allow, the dissident Bushmen another sip of water.

I approached the threshold to the central Kalahari still suffering from the delusion that the stronger and more durable civilization was the outside one surrounding the reserve, and that it was I, the big-hearted, well-situated, and self-righteous American, who might somehow come to the rescue of those last, poor, helpless Bushmen.

I LURCHED UP to the Kalahari reserve's entry gate and killed the engine. As it cooled, my heart beat faster, and I felt my shirt peel off the seat, sticky with sweat. I began to grow anxious that the officials would keep me out or inspect the vehicle and uncover the "contraband," perhaps leading to my arrest or deportation, or worse. I had not set out to become an amateur water smuggler, and, looking back, I could not explain my rash decision to sneak supplies into defiant strangers living a raw desert existence. Except I felt I had no choice.

As it turned out my anxiety was somewhat misplaced. The official was neither sinister or menacing, just bored. After I signed some tourist forms, paid a fee, wrote down my license plate number, and lied about my intended destination, he waved me past and into the reserve. I revved up the engine, leaned back, and breathed deep. As I shifted into first gear I chuckled, imagining that my most dangerous obstacle had been Botswana officialdom and that my worst ordeal now lay behind me.

Late in the afternoon of the following day, somewhere north of the Tropic of Capricorn, my engine suddenly locked up and gravity took over. In less than a second all four wheels pressed down beneath one ton of steel, and forward motion ceased. Since the heat and fuel gauges were normal, I held my breath and tried the ignition. It rolled over and over and over and over as the electric pulse weakened each rotation, until in a panicky shudder I jerked out the key and stepped out into the blast furnace.

A look under the hood diagnosed the source of my predicament. The air hose had rattled loose some miles back, inhaling unfiltered dust. With each artificial breath, tiny airborne particles had mixed with gas vapor to accumulate inside until the carburetor clogged. Starved of oxygen, internal combustion suffocated and all eight pistons froze.

The vast landlocked beach yawned out in every direction and absorbed my screams. But gradually, over the next few hours, this desert began to seem less hostile than merely indifferent. Animal sounds reclaimed the dead silence as an insect hum built to crescendo, songbirds chir-chir-chirred in dry leaves, and barking geckos cracked their throats. Out here was no good or evil, only thirst. All life sought moisture from tiny pockets between grains of endless sand, from the sap of plants and from blood.

Stupidly, I had been driving by myself. Worse, to avoid arrest or imprisonment, I had misled the country's federal officials about my true destination and smuggling operation. At the time I had thought my tourist ruse clever, but now no patrols would come searching, not here. Retracing my tracks on foot would be impossible, for I'd have to carry eighty pounds of water on my back. I had come too far, too fast, too deep into the reserve, and during that first day out there was already sweating three pounds a day by just walking slowly in circles. Below a rationed intake of water, my cerebral tissues would desiccate like a sponge in an oven. So as the sun governed the daylight, and lions owned the dark, I lay on the vehicle roof tent anticipating a quiet, slow death here, wondering what it might feel like.

My girlfriend had helpfully provided me with a paperback by an adventure writer who narrated all the ways careless thrill seekers can meet their end; the last chapter, on dehydration, defined progressive stages of "desert thirst" in such cheerful terms as "clamorous," "shriveled-tongue," "blood sweat," and "living death."[6] It then helpfully related the classic story of W. J. McGee's description of a prospector caught for days in Arizona's Gila Desert without water. McGee encountered "the wreck of Pablo," with a weak pulse, barely alive. He was stark naked, having maniacally shed all clothing, food, and possessions, wandering in a hallucinatory state until collapsing nearly deaf and blind. "His lips had disappeared as if amputated, leaving low edges of blackened tissue; his teeth and gums projected like those of a skinned animal, but flesh was black and dry as a hank of jerky; his nose was withered and shrunken to half its length, the nostril-lining showing black; his eyes were set in winkless stare . . . The mucus membrane lining mouth and throat was shriveled, cracked, and blackened, and his tongue shrunken to a mere bunch of black integument. His respiration was slow, spasmodic, and accompanied by a deep guttural moaning."[7]

So I had that to look forward to.

By that night at four A.M. came a low rumbling in the distance, and within seconds I was off the roof, down on the ground barefoot racing diagonally toward the distant headlights of the deus ex machina that was a group of well-equipped tourists headed toward wildlife pans in another direction. Armed with better tools and know-how, the white South Africans opened my engine's valves to suck through the dust clog. The engine roared. I inhaled relief and exhaled gratitude and gave them all my wine.

"Ag, you're a lucky man," one man said. "Traveling alone like this."

"You mean stupid man. I know. I owe you more than you can imagine."

"If you don't mind our asking, um, what the hell brought an American out here headed off in this direction?"

I explained I was researching the causes and consequences of water scarcity, and exploring Botswana as a crucible for adapting to coming droughts worldwide.

Another squinted at me like a doctor scanning a sick patient. "You know," he deadpanned, "there might be easier ways to find out than this."

I nodded, attempting a broken smile.

"You ought to come follow us. Stick with our convoy."

Their offer was tempting. I had been stranded for a mere twenty hours, but it felt like twenty days. Safety in numbers reduced risks, and loneliness. I could join a relaxing wildlife safari and escape back to civilization, where I might one day look back and laugh at my misguided rescue mission, my crusade to the Bushmen, punctuated by idiocy. Only a delusional fool would politely decline.

I politely declined. But in my defense I figured it was long past time to confront on its own terms that third inevitable, after death and taxes, which was heat. Rising dry heat was sucking moisture from soils everywhere. Drylands—stretching across two fifths of the planet's surface and holding a third of its population—already felt the warm breath that melted away the earth's storehouses of snow and ice, inhaled anemic rivers, and evaporated the drinking water reservoirs that supplied Sydney, Mexico City, Jerusalem, Beijing, Tehran, or New Delhi. I could run away, somewhere, anywhere. But hot dry air would follow and catch up. There was, as far as scientists could see, no escape.

My desert trial alone in the wilderness had awakened in me no comforting spiritual conversion. I remained a secular humanist with elitist prejudices. Facing a malignant tumor, I'd check into the Mayo Clinic; if indicted, I'd phone the most ruthless defense lawyer; and now, as my crowded planet was diagnosed with a terminally warmer future, I'd seek out the most experienced survivors of perpetual drought and water scarcity.

I waved off the kind and heroic tourists and rolled on in the opposite direction, having experienced a conversion of another sort. It dawned on me that it was no longer I who could help these Bushmen endure the hard times that had been thrust upon them, but rather they who might guide us through the coming Dry Age of our own making.

Over those few days I had but glimpsed the potential horror humankind faces at the end of water; the native people out here lived that dry existence by choice. Defying official orders, the last few bands were said to roam just the other side of the horizon. So while nervously checking that engine air hose every hour, I lurched over the maddening sands into the arid savanna where man was born and where his scruffy unwashed descendants still gave birth and teased each other and smoked and hunted and danced, and died.

CHAPTER 3

Intransigent Eve

FOR MILLENNIA BUSHMEN ROAMED THE African subcontinent exclusively, with only the survival of their kin to ensure. But eventually the arrival of waves from the pastoral, agricultural, and industrialized world would constrict inward on the central Kalahari and impose on them an additional burden. Some of the initial relationships held out hope for peaceful coexistence, as Bushmen taught newcomers how to find, conserve, and live within the confines of scarce water. Still, many of these better-armed thirsty outsiders would be caught unprepared for the sudden onset of dry conditions.

In 1890, with European powers embroiled in the "Scramble for Africa," imperial England launched a bold venture to colonize the subcontinent's dry heart. The undertaking was the brainchild of Cecil John Rhodes, Queen Victoria's favorite, and his design was to preemptively expand Britain's sphere of influence by blocking incursions by Germany, Portugal, and Belgium. Success or failure would have geopolitical repercussions.

Rhodes offered guns, provisions, financing, and property title to eighteen families if they would settle at Ghanzi Springs in the western Kalahari Desert. But in order to claim and cultivate that promised land, the party had first to cross the great sand sea. The group planned as well as it could, but soon discovered the sand to be softer, hotter, and drier than expected, and their slow, plodding livestock to be woefully ill-adapted. The advance party's supplies quickly dwindled, and, alarmingly, they found precious little surface water at the so-called springs. Drought overtook them. As provisions declined and cask water shrank, the leaders faced a wrenching dilemma: stick together to die slowly from hunger and thirst, or leave families behind as men made a mad dash for support and rescue.

One relationship tipped the scales. The indigenous people in that dry region had shown themselves without hostility or tension and displayed an astonishingly intimate understanding of the desert. So after much agonizing and hasty embraces, husbands and fathers entrusted wives and young children to local bands of Bushmen and raced back across the yawning Kalahari.

The drought worsened. Weeks became months, and the families lacked any means of communication. Nearly two years passed before the rescuers could return, and when at last the men secured funds and provisions and transport back across the desert to their loved ones, they did not know who, if any, might still be surviving on the far side.[1]

At that time almost nothing was known about Bushmen. Whenever encountered by European settlers in southern Africa, the indigenous bands became subjects of brief curiosity, often swiftly overcome by relentless brutality. As in America and Australia, encroaching farmers and herders saw native inhabitants as less than human, best shot quickly as vermin. As conflict over arable land and water arose, the outsiders enslaved, exploited, abused, and exterminated any defiant Bushmen in their path, leaving behind only a one-sided rationale as a record to justify their deed.[2]

Given a few years, or better provisions, perhaps Rhodes's settlers might have done the same thing. Instead they had been desperate enough to throw their families on the mercy of these diminutive Kalahari strangers, and as it turned out, they could not have trusted more capable people. If anyone knew how to endure dry times, it was Bushmen.

When the men returned, they discovered to their joy that, despite the severity of the drought, the Bushmen had taken excellent care of their women and children. Somehow Bushmen squeezed enough water and food from the land to quench the thirst not only of their own families but also of those sent into the wilderness by Queen Victoria. The men found their families to be healthy, happy, and strong.

Given our current predicament of drought and water scarcity, it is worth asking how Bushmen could thrive under such unforgiving conditions. Fortunately, a rich archive of material can be mined for answers; in the last six decades few ethnic minorities have been more exhaustively probed. And among all the professional scholars who came and went, one American dynasty studied Kalahari Bushmen for a long time and in great depth and detail.

In the 1950s Laurence Marshall, the retired Raytheon founder, holidayed with his family in southern Africa's interior. Over the next five decades his wife (Lorna), son (John), and daughter (Elizabeth) would document the Bushmen way of life in classic works: books, essays, photographs, and films. The family's last survivor, Elizabeth Marshall Thomas, revealed how the old women were never ignored by the young or the men because the females in particular ensured a viable approach to the constant need for water. "The older someone is, the more that person remembers about what happened before the rest of the group was born, events that, without written records, would be lost if someone couldn't describe them," she explained in her beautifully written and evocative memoir, *The Old Way*. "In the event of a fifty-year drought, for instance, it would be those in their sixties and seventies who might remember a way of getting water, or of getting by without water, something that their own grandparents had shown them the last time this occurred."[3]

It was most likely these older matriarchs who ensured that Rhodes's first white settlers had enough water, when women and children were entrusted to them, and who thus helped forge between Bushmen and whites a long-term sense of reciprocity. For decades afterward and up to the present, a custom was passed down among those original families. To honor their eternal obligation to Bushmen, the remaining descendants of Rhodes's advance party would never refuse any destitute Bushman's desperate plea for food, shelter, living space, or water. It was payback of debt, an African equivalent to America's own tradition—in theory if seldom in practice—of Thanksgiving.

But in a land of perpetual scarcity, tensions ran high, and memories gradually fell short. Newcomers saw customary reciprocity as a silly sentimental nuisance. If you provided water to a landless Bushman, said critics, he would, like any human, begin to think of that water as his entitlement, his fundamental human right. He became dependent on you to slake his thirst, which could only grow. He did not value the water, if freely given, but would in fact surely waste it. No, if you took responsibility for providing water to anyone, you took responsibility for his life.

British colonial officer George Silberbauer understood both perspectives. From his landmark Bushmen surveys in the central Kalahari during the 1950s, he valued water as utterly precious, as his "radiators erupted with the regularity and violence of geysers. At one stage we were so short of water that we had to try to make do on about 5 liters each per day." He knew Bushmen made do on far less, yet stuck to his decision that seemed cold yet in retrospect, prophetic:

> I made it clear from the start that I would not share our water with the Bushmen. To have done so would have reduced the survey to a water-carrying operation and might have made it difficult for the Bushmen to re-accustom themselves to going without water when we were no longer there. I always felt unhappy about the disparity in our circumstances, but the Bushmen did not seem to resent it.[4]

Subsequent travelers lacked his foresight. Several decades later, Botswana initiated water-carrying operations throughout the country, altering the delicate balance of political power and economic negotiation in the Kalahari. And that, as the Marshall family and some anthropologists concluded, brought the beginning of the end. Bushmen grew dependent on cattle posts, on do-gooders, and on governments to provide their water. In the process they grew dependent on their own Gods-Must-Be-Crazy appeal among romantics who lived outside the Kalahari, becoming victims of what the late John Marshall christened "Death by Myth."[5]

NEAR CAPE TOWN some part-time Bushmen dressed up for tourists and danced around fires. In Lesotho a few pretended to paint on caves with techniques and pigment that may or may not have in any way resembled their ancestors. Trackers in Namibia's eastern Bushmanland helped foreign great white hunters kill bull elephants. The apartheid-era armies gave Bushmen in Angola food, money, guns, and instructions to use their hunting skills to track the communist enemy and then later cast them aside once the cold war ended. As one southern African indigenous leader lamented, "Governments want to control us, missionaries want our souls, and environmental organizations want our resources and our support."[6] Bushmen became an "African underclass," a status that fueled intense global debate about the Bushmen's true identity.[7] A new school of revisionists began to accuse "essentialists" of idealistic prejudices, which wrongly forced Bushmen into a false past and, therefore, false present. Not only did no "true, authentic Bushmen" still remain, went the new argument, but they never had really existed in the first place quite as isolated and exceptional as previously claimed.[8]

The so-called Great Kalahari Debate split open the academic discipline of modern anthropology until it seemed no external evidence could resolve the dispute or reconcile the warring antagonists.[9] Then around the turn of the twenty-first century, from the cutting-edge science of population genetics, came an unexpected source of internal evidence.

Affordable computerized data systems of biotechnology could break down, sequence, and unlock the secret code retained in the molecules of human cells—specifically cells embedded within female mitochondrial DNA and male Y chromosomal DNA. Genetic mapping projects began to test the bloodlines of global residents for definitive scientific links, seeking mutations—the spelling errors found in all "genetic markers"—to trace our ancestry.[10] Once testing began, American researchers could reel back through time, tracking who begat whom, racing along family tree tops and branches to family tree trunks to our deepest blood roots predating the colonial or pre-Columbian eras, right down to our earliest human origins.

Indeed while genetic DNA testing couldn't "stop," it could and eventually did reveal, with remarkable precision, a human biological "start." Native Americans could trace ancestral origins west and north to North America's first humans arriving 15,000 years ago across the Bering Strait, where Asia itself was settled 50,000 years ago. Caucasian settlers could trace their roots to 35,000 years ago in Europe, and black descendants of slaves discovered in Africa the richest and oldest genetic diversity of all.[11] No one in the New World, Europe, Australia, or Asia could claim to be "indigenous." Under the skin our ancestors all arrived as immigrants; hence every bit of DNA in every human traced back unequivocally to a singular source in Africa.[12]

Then it got even more interesting. While tracing our earliest roots, geneticists discovered a point in Africa where early *Homo sapiens* almost went extinct. A cataclysmic drought had swept across the continent, wiping out humanity until scarcely a few thousand of our species endured.[13]

Another megadrought had scattered off many of our human ancestors to populate the earth in a mass exodus. Only the hardiest survivors remained in small, isolated pockets on the continent of Africa.

By sifting through that shrinking African pool of diverse DNA samples, researchers traced back 150,000 years until a single female bloodline emerged from a primordial Eden. From that anthropological Eve, all humanity had descended, over two thousand generations, to the 6.1 billion of us alive today.

The longest strain traced down from the earliest drought survivors is today collectively shared most deeply by those diverse bands of people who still live in the Kalahari Desert. So among living Bushmen, DNA scientists affirmed how the elder matriarchs among them accurately portray a modern genetic reflection of Eve.[14]

Apparently Eve existed, but she was not blonde and fair-skinned, plucked from Adam's rib. Instead she more closely resembled a savvy, bawdy, wise, and wrinkled light red-skinned forager, dancing and hunting and healing alongside her children and grandchildren in the Kalahari, all of them living proof that her bloodline knew how to survive, to adapt, and to overcome Earth's deadliest climatic force.[15]

Yet by 2002, just as that force began to press down on us in earnest, it seemed too late to learn much from those Bushmen who remained in any real or imagined Eden. Elizabeth Marshall Thomas said her former harmless people had forgotten what it meant to be self-sufficient in an arid land. No one alive in Africa today still lived the life of a hunter-gatherer, she wrote. No one could teach the next generation how to cope with the coming Dry Age. Those old matriarchs who once knew the ways of water were either long dead or living on handouts. "If you happen to see a contemporary film or photo showing Bushmen dressed in skins, perhaps beside a small grass shelter or following a line of antelope footprints or handling a bow and arrow, you are seeing a reenactment," she lamented. "Today, nobody lives in the Old Way."[16]

Perhaps she was right, and the dream was gone. But then, perhaps Bushmen could prove how complex, adaptable, and resilient they are and could reclaim their survival strategies. Perhaps evolution never entirely froze at some past age or era, and old ways of coping with water scarcity could be recalled again and again under duress in the face of necessity.

At the time this book went to press, one free and newly autonomous Bushmen society still clung fast to the threads of its thirty-thousand-year-old existence. These Kalahari dissidents now endured within—yet independent from—Botswana. Indeed, not only did Bushmen lack so-called government help; they faced officials' direct, acute, and relentless hostility.

Unlike the biblical Eve, their matriarch knew all too well the distinct ways and tracks of the serpent, */xaudzi*. She avoided venomous snakes and sought indigenous medicine to treat bites; others she killed and ate. She had not sinned in listening to the knowledge of which wild food to taste. She did not wish to depart, go off, till the earth, and feel shame. With a fierce sense of pride she loved her home, that dry thorny sand-filled Garden from which she and her mate had been formed. As the voice of power issued orders to banish her forever, she quietly disobeyed.

CHAPTER 4

The Desiccation of Eden

ON MY FIRST ILL-PREPARED JOURNEY INTO the desert I got hopelessly disoriented, and while bleeding gasoline from leaky plastic fuel canisters I began chasing my own wheel tracks backward for hours. As the deep sands repeatedly trapped my wheels I grew grimy, gritty, thorn-scratched, engine-oiled, bee-stung, and tooth-rattled. The scenery droned past, and my radio played nothing but static. Right from the start, even before the pistons locked up, I knew I hated this desert. I loathed it. Finally, days after my rescue by tourists and on completion of the smuggling run, as my wheels once again rolled over smooth asphalt, I heaved an inward sigh of relief that burst out into giddy laughter, realizing I had utterly lucked out, narrowly escaping from what would have been a genuinely stupid death, and so let me tell you, I was through, done, gone, finished, out of there and swearing out loud to myself how absolutely nothing on earth could ever, ever lure me back into that wretched monotonous hellhole again.

I returned more than a dozen times. Something kept pulling me back. "It's a strange thing, the Kalahari," agreed Arthur Albertson, who had mapped Bushmen territories in the reserve on horseback with GPS, occasionally stalked by lions. "Gets under your skin, doesn't it?"

It did. And while it was impossible to explain its draw to my friends and family, there it remained. Much of its spell was the people, of course. But even beyond them the Kalahari is so haunting because its immensity feels, like outer space, so utterly indifferent to life.

Over the eons, winds had blown coarse grains of sand to blanket portions of nine nations. From a narrow wedge over Gabon in the northwest, the sand mantle cascaded down across northern South Africa. Across the Kalahari stretched the most extensive unbroken expanse of sand on the face of the earth.[1]

That said, such "sandscapes" are no longer confined to the Kalahari; in recent decades they have begun to spread faster and almost everywhere. Soils degrade under intense pressure combined with hotter, drier air currents. Abusive and careless land use cause the intricate networks and crusts of lichens, mosses, and bacteria to deteriorate, rending the organic fabric that binds mineral particles together. Arid deserts and semi-arid savannas have expanded across one hundred countries—creeping especially across Africa and Asia, but also American states as diverse as Maine and New Mexico and, as witnessed in the 1930s "Dust Bowl," the Midwest.[2] With rising heat and windstorms, sands cross borders, even oceans, scattering nutrients with them to lay a thin infertile blanket over new lands. The Sahara keeps moving south at thirty miles per year.[3] Vast swaths of Africa's degraded expanding Sahelian region wind up landing, useless, in southern Europe. Each year, China loses a Rhode Island–sized parcel of fertile land to desert; a fraction of its ensuing sand and dust storms cascade over North America.[4] But at least half of the planet's dust in the air today has come from arid Africa, and the impact of the drying, everywhere, means the atmospheric dust has increased over the last century by a third.[5] No need to venture into the inhospitable Kalahari; the deserts are coming to us.

While not as acutely dry as the Namib, Mojave, Atacama, Gobi, or Sahara, the Kalahari's heart lacks moving or standing water. Some visitors say its inward-turned, desert-terminus, shallow pan dynamics resemble a flatter, drier, sandier version of the Great Basin in the United States. Also like America's western region, the Kalahari's outer edge is fringed by rivers—the ephemeral Limpopo, Molopo, Nossob, and Boteti. Few of these streams flow reliably, if at all, and within the arid heartland are only fossil valleys, broad desolate depressions, reminders of where greenery once thrived but no longer can. In the north the Okavango and Kwando stop short of the vast, flat empty basin, known to locals as "the place where rivers go to die."

Botswana has long been prone to low humidity, high evaporation rates, fickle clouds that might dump three inches of rain in forty minutes during magnificently violent thunderstorms, or nothing at all. But now aridity keeps getting worse. During Qoroxloo's lifetime rainfall in the tropics already had declined a fifth while evaporation temperatures rose two degrees, and now climate models fix southern Africa as the continent's epicenter of drought. Within decades another expected 20 percent drop in precipitation is predicted to make the region "completely dry up."[6]

IF DESERTIFICATION COULD be confined to Botswana, that unhappy fate might be sad for Botswanans but comforting to those of us living a safe distance away. Unfortunately, climate scientists say this Kalahari scenario—hotter, drier and longer droughts punctuated by increasingly rare sudden deluges that can't easily be contained—appears to be coming soon to a landscape near you.

The spread of aridity initially confounded early climate models; a warmer world should at least in theory lead to a wetter world. Yet for many dense populations the reverse was unfolding. Due to the heat, the proportion of the planet's land surface suffering drought in recent decades had doubled.[7] From Eurasia and Australia to the Americas and Africa, tropical regions were experiencing dry conditions that had not been seen for the last seven hundred years. Scientists confirmed that no place was safe. Climate change had begun to reveal its ability to spawn megadroughts anywhere on the planet.[8]

What's more, it turns out that the self-proclaimed climate skeptics are right to incessantly remind us how global warming is as old as time and has produced winners and losers. But the less comforting wrinkle they leave out is how, time after time, whenever climate changed in the past, North America lost.

New hard evidence, accumulated from tree ring data and pollen counts, suggests that devastating droughts have shattered human settlements, dating back to when people first arrived in North America.[9] Paleoclimatology remains a young and inexact science, and no one can pinpoint the precise stages at which high temperatures and dryness caused local human extinctions. But the correlation is sobering. Wherever scholars searched, further and farther, they discovered how the sudden change from a wet to dry climate caused most extinctions of flora and fauna—the living habitat on which humans depended. Unusually severe and protracted drought ranked highest among the most devastating and calamitous of all climate events because the resulting water scarcity brought wildfires, crop failures, livestock deaths, food shortages, and famine.[10]

At the end of the Pleistocene era, temperatures on the continent warmed by roughly thirteen degrees Fahrenheit; with less surface moisture to moderate, cold winters grew frigid and hot summers fried. The climate transformed entire forests and withered grasslands, wiping out or driving off the prey base upon which humans fed.[11] Some five thousand years ago, flourishing Native American cultures suffered prolonged exposure to climate only slightly hotter than it is today and nearly went extinct; "for more than a millennium the southwest was little more than one big ghost town."[12] A hot era that lasted from 800 to 1300 boosted medieval European agriculture but scorched much of pre-Columbian America.[13] In Northern California, the key acorn harvests failed, dispersing that entire region, while a Southwestern megadrought snuffed the Anasazi civilization.[14] North America's most complex pre-Columbian desert civilization, the Hohokam, lived for a millennium on the banks of the Salt and Gila rivers of what is now central Arizona. Backed by a canal system that rivaled the Romans' in scale, scope, and execution, they thrived in a landscape that remained often hotter than one hundred degrees Fahrenheit, relieved, sometimes, by only seven inches of rain. But recent pollen sequence analysis revealed that during that medieval warming period, as three of every five years turned dry, the Hohokam peoples were obliterated.[15]

Prehistoric collapse in the New World might be dismissed as the weakness of "primitive" cultures who would be overrun by dominant European newcomers equipped with the proverbial guns, germs, and steel to endure. But again, new evidence reveals that despite superior technology, immunity, and weaponry, America's first colonies were, if anything, less adept at coping with protracted thirst. Queen Elizabeth's first settlers at Roanoke were last seen on August 22, 1587, hungry and running out of water, during a dry spell so severe that it even affected the native subsistence food of indigenous Croatoan tribes upon whom the colonists depended. Three years later they had vanished. Following centuries of mystery, a recent tree ring reconstruction from A.D. 1185 to 1984 showed that the Lost Colony precariously arrived at the onset of the region's driest three-year episode of the last eight centuries.[16] Two decades later, forty-eight hundred out of six thousand Jamestown colonists died in waves upon their arrival. Early historians blamed the deaths on dumb planning, incompetence, and weak support, but scientists have now directly and precisely linked the sudden crash—in native subsistence, peak mortality, domestic livestock deaths, and a rapid decline in drinking water—to the driest seven-year period in 770 years. Unlike the colonists at Roanoke, these settlers left written records of what occurred. As water dried up, Jamestown's former "London Gentlemen" degenerated into thirst-wracked, scurvy-ridden starving wretches turning on each other, killing and even eating members of their own family.[17]

Following those first unfortunate colonies, the geographically blessed United States enjoyed an exceptionally cool, wet era during which it progressed from agricultural and mercantile economies to a postindustrial Information Age of 300 million highly urbanized people. Even so, during the wettest century of the past millennium a few dry "speed bumps" have profoundly destabilized Americans, suggesting the level of risk water scarcity holds in store for us. A relatively mild six-year drought in the 1930s wreaked agricultural and social mayhem throughout the Dust Bowl. A less acute but more widespread drought pressed down across the Midwest during the 1950s, extinguishing many rural economies. Over subsequent decades the already arid Southwest and West grew increasingly dry.[18] Starting this century, even laypersons across America have been observing everyday weather that seems hotter and drier than normal.

Scientists confirm that in fact conditions are extreme, and climate will likely worsen our stability in the decades ahead. As we humans burned and cleared vast forests, converted land to irrigation agriculture, and powered industrial growth with fossil fuels, we were unwittingly baking the earth in what appeared to be an irreversible process. Our carbon emissions had thickened the relatively thin layers of the outer atmosphere, trapping solar radiation. The effect resembles leaving our collective car in an exposed parking lot with windows sealed and kids locked inside.

One troubling indicator is revealed by our jet stream—the horizontal conveyor belt of cold air that dips down from Canada across the United States and creates national weather systems. Recently scientists found that, by conservative estimates, it has been migrating northward at a minimum of 12.5 miles per decade, or eighteen feet per day. As it moved up from a region, high pressure and clear skies converged in its wake, leaving the South and Southwest hotter and drier. Animals may keep up with that pace; habitat cannot. "Look south of where you are," said Ken Caldeira, a climatologist from the Carnegie Institution, "and that's probably a good idea of what your weather may be like in a few decades."[19]

Or sooner. A consensus of climate scientists agree that under current arid trends, the Colorado River will continue dropping to half its current flow, and stay there, even as the urban population that depends on it for food, water, and energy continues to double every few decades.[20] One recent book predicts wholesale collapse and mass exodus.[21] The West has always been dry. But now it seems droughts have gone national, hammering even formerly lush regions like the Southeast.[22]

In 2007, not far from where the Roanoke colonists vanished in 1587, in today's North Carolina, the National Climatic Center recorded that six of America's ten warmest years had occurred since 1998, rising 0.6° F for each of the last three decades. That same year the state endured the driest year in recorded history,[23] and by November, seventeen water systems had one hundred days of water supply remaining before they reached rock bottom.[24] Farmers hauled water by pickup, while Raleigh and Durham nearly dried up after reservoir levels plunged.

As streams fell beneath historic levels, low water translated into low energy. The state's nuclear reactors and other power plants were crippled, throttled back, or temporarily shut down since the billions of gallons of water needed to cool them had evaporated.[25] In the drought zone, as millions of customers braced to pay ten times more or lose their power,[26] the state turned in vain to southeastern neighbors for help. Extreme drought cut Tennessee Valley Authority hydropower in half, exposed Lake Okeechobee's bare muddy bottom, dried up $787 million of Georgia's crops, and left Atlanta—America's fastest-growing city—with sixty days of water. Under cloudless skies the nation's biggest urban agriculture industry laid off thirty-five thousand workers, its massive Stone Mountain Park melted 1.2 million gallons of manufactured snow, the earth's largest aquarium drained new exhibits, and the Coca-Cola Company ominously shut off the fountains at its corporate headquarters.[27]

The dark side of drought goes beyond economic stagnation to cause political implosion.[28] Just as it had shattered the Roanoke and Jamestown colonies, scarce water fragmented society, curtailed liberty, and eroded trust. Government rationed individual water consumption to one tenth of what we normally consume each day.[29] It cracked down on private well pumps, claimed and regulated them for public consumption. "Other hazards tend to pull people together," said Michael Hayes, director of the National Drought Mitigation Center, speaking of water's power. "With a drought, because it's a limited resource, it tends to drive people apart."[30]

Divide us it did. Southeastern states have sued one another for remnant water, and even Maryland challenged Virginia over control of Potomac River currents for the first time since the Civil War.[31] As citizens appealed to government, governors appealed to God. In 2007 Alabama governor Bob Riley declared a week in July "Days of Prayer for Rain." In November, Georgia governor Sonny Perdue gathered people on the capitol steps, bowed his head, and appealed to a higher power for relief. "We've come together here simply for one reason and one reason only," he told the gathering, "to very reverently and respectfully pray up a storm."[32]

Irreversibly rising heat, migrating jet stream, booming industry, thirsty populations, helpless leaders: the Perfect Drought.

It doesn't appear to be getting any cooler or damper; both the World Meteorological Organization and the British Meteorological Office confirm that the last decade marked the hottest on record,[33] and reputable observers maintain that our current megadroughts represent the overture of what will follow for centuries. Based on new evidence that the Global Warming era was dawning sooner than expected, even global warming guru Al Gore changed his mind: Prevention alone was not enough, he said. In the eleventh hour on our warming, drying planet we must rapidly learn to adapt.

But then, who will teach us how to cope? For the last seven years as the United States broke records for high heat and low reservoirs and prepared for the worst hot Dry Age in thirty thousand years, the remnants of the world's oldest civilization—the only people with the survival experience, strategies, tactics, and values to guide us through the extremes of our once and future drought—were embattled in the heart of the Kalahari Desert, surrounded by armed men who were urging these last free Bushmen to surrender their way of life forever.

CHAPTER 5

Besieged and Besieger

BOTSWANA'S USE OF THIRST AND WATER to coerce insubordinate people at first seemed extreme and archaic, but on closer inspection the tactic was neither. Modern governments have routinely deployed water as a weapon, sometimes rather casually. Perhaps what most set Botswana apart were the disproportionate stakes, scope and scale of its ambition.

At least in the beginning, there were believed to be between seventeen and eighty-nine defiant Bushmen. To put that in perspective, an arid landscape larger than New Jersey and Massachusetts combined was holding fewer people than the Texas Alamo. Surrounding them, President Mogae set up an invisible cordon and reinforced the perimeter of the reserve. He began to close off access, regulating all who entered or exited, under the tranquil outward appearance of being calmly at peace. Within months his armed officials constricted inward toward the Bushmen camps, turning the reserve into the twenty-first century's largest and longest-lasting siege.

By definition a siege occurs whenever a dominant outside force encounters and surrounds people who, despite entreaties, refuse to submit and who cannot, for various reasons, be taken by force. Siege has three tactical dimensions: isolate the target population; block the reinforcement of provisions, especially water; use deception to bypass defenses. Year after year, Botswana deployed all three against Bushmen.

Among the more lopsided aspects of the Kalahari siege was the rank of those who ran it. The armed and uniformed officials who surrounded

Qoroxloo's homeland were mostly game wardens, guards, wildlife patrols, district administrators, and a police squad. But they weren't calling most of the shots—the national security apparatus was. The siege was ordered, orchestrated, and enforced by Botswana's military and diplomatic corps.

In an arid land, water is a matter of national defense. So perhaps a forceful response seemed necessary to maintain order, even martial order. Behind the technocratic president Mogae stood the vice president, ex-general Ian Khama; Major General Moeng Pheto; and the foreign affairs minister, Lieutenant General Mompati S. Merafhe. This contingent saw political matters through a security lens, and the Bushmen siege was no exception.[1] When asked about cutting off water and the siege, General Merafhe said he was "proud" and had "no regrets." He explained, "we can't allow them to live that life of keeping on running after wildlife as a means of survival."[2] Yet in a testy exchange Merafhe blurted out the gaffe, a statement in which a politician tells the truth: "We put these people . . . where we want them to be."[3]

These people were unarmed, with no fortifications. They posed no threat and never counterattacked. Their allies were few, distant, and dispersed. Rather than punishment, Bushmen faced generous rewards of food money and water if they surrendered. Yet for reasons Botswanan officials could not fathom, Bushmen refused to yield, and when approached, they melted back into the arid sandy bush.

ANALYSTS CALL SIEGE "the oldest form of total war"[4] and say it represents the ultimate meltdown in peaceful relations among equal forces. Botswana's president did not consider Bushmen like Qoroxloo his equal; and he bristled at any mention of force. He maintained throughout his tenure that the country was still at peace in the Kalahari, that those dissident Bushmen who refused to leave their homes were not enemies to be beaten, merely poor and misguided creatures to be pitied. Claiming to be acting in their own best interests, President Mogae drew a line in the sand around the reserve—unlimited water outside, unlimited thirst inside—then sat back and waited. Advisers assured him Bushmen could not last long. They would quickly succumb and soon request, if not beg, for transport to depart to a resettlement zone where taps ran plentifully.

Siege had defined geopolitical struggles long before Homer's Muse sang of Troy. The tactic brought horrors and risks for both sides but often proved decisive. Union sieges of Vicksburg and Petersburg telescoped the closure of America's Civil War, just as the Spanish Civil War accelerated through the sieges of Toledo, Gijon, and Madrid. World War II pivoted largely on the Nazi's failed but lethal siege of Leningrad, which left one million dead. France left Vietnam after Viet Minh forces laid siege to Dien Bien Phu. And so on. Yet if wars often hinged on sieges, sieges turned largely on control of fresh water.[5]

Caesar took the besieged citidel of Uxellodunum after ordering his archers to shoot water carriers who drew from the river; his engineers then diverted the last remaining spring. Two thousand years later in Bosnia, little had changed. Serb forces cut power to pumps that delivered water, while in the ensuing siege of Sarajevo, gunners and snipers picked off 90 percent of its thirsty civilians who tried to fill plastic containers at rivers or water points miles from home.[6] Even in police sieges the decisive element was water.[7] At its birth, Israel endured sieges by Palestinian Arabs followed by Jordanian and Egyptian forces;[8] six decades later the tables had turned, as Israel besieged Gaza, cutting off the supplies that inhabitants needed to pump drinking water.[9] Perhaps the most pivotal siege event in North America took place in 1664, on Manhattan Island, when British warships sailed up the Hudson River and drove the residents of what was then New Amsterdam to retreat behind the walls of a fort that contained no water sources. Crippled by thirst, the ensuing surrender of the besieged Dutch was as swift as was the installation of public wells in the newly renamed New Yorke.[10]

Today, however, the West condemns any siege of peaceful residents. To protect innocent victims from conflicts, Article 54 of the 1979 Geneva Convention explicitly forbade nations:

> to attack, destroy, remove, or render useless objects indispensable to the survival of the civilian population, such as foodstuffs, crops, livestock, drinking water installations and supplies, and irrigation works, for the specific purpose of denying them for their sustenance value to the civilian population or to the adverse Party, whatever the motive, whether in order to starve out civilians, to cause them to move away, or for any other motive."

Most similar actions brought sanctions. And indeed, as the Kalahari siege ground on, U.S. diplomats repeatedly rebuked Botswana for its tactics against Bushmen.[11] Then again, America lacked moral authority just then, having proven itself all too willing to launch a large-scale siege of a vast and arid land. The Pentagon was not innocent of waging thirst against trapped civilians.

In 1995 the U.S. Defense Intelligence Agency declassified documents drafted four years earlier. These described vulnerabilities of Iraqi water resources in exquisite detail. Like Botswana, Iraq was a dry country; Agency documents observed that Iraq "could not rely on rain to supply adequate pure water." Like Botswana, Iraq had to import all parts and chemicals like chlorine to purify its brackish, polluted, and mineralized water supply. A few elite families could buy and truck personal water from the north and boil it for consumption. But the country's vast majority lacked enough pipes or money to meet basic needs.

The agency then spelled out what would happen after America destroyed Iraq's clean water and blockaded critical resupplies under an international sanction. For starters, "epidemics will become probable."[12] Within three to six months, specified the documents, the loss of clean water will drive up incidences of "acute respiratory illnesses, typhoid, hepatitis A, measles, diphtheria, pertussis, meningitis, and cholera." The agency itemized further likely outbreaks of acute and fatal diarrhea brought on by *E. coli*, shigella, and salmonella bacteria, giardia protozoa, and rotavirus, all of which will affect "particularly children" (a phrase the documents put in parentheses). In short, the U.S. government—under the elder Bush and Clinton administrations—used sanctions against Iraq as a calculated form of modern siege. It tightened the screws to degrade the country's water supply, and it did so with the knowledge of exactly how and how many thirsty Iraqi civilians "(mostly children)" would sicken, suffer, and die.

Throughout the 1990s, the U.S. policy was to deliberately destroy water treatment. During that time, UNICEF estimated five hundred thousand Iraqis (particularly children) died as a direct result of sanctions on water treatment chemicals, filter membranes, and tools. Thomas J. Nagy, the George Washington University business professor who first brought the declassified DIA documents to light, published his findings in early September 2001, concluding, "No one can say that the United States didn't know what it was doing."[13]

There were obvious political benefits to this passive and indirect tactic. Siege had the advantage of appearing more effective, more diplomatic, and less visible than noisy firepower. Isolation in a waterless siege meant hands remained clean. Civilians killed indirectly from thirst could not point to a blast or bullet; they were even blamed for their own thirst.[14] Siege was the difference between blatantly poaching the last endangered fish to extinction, and quietly draining their lake.

Drying up the enemy was a powerful weapon. Failing a military outcome, sieges were almost always decided when people succumbed to hunger, disease, or acute thirst. The goal of a siege was attrition. Defenders and civilians were reduced to eating anything vaguely edible: horses, dogs, shoe leather, or each other. On occasion, the besieged would drive surplus civilians out to reduce the demands on stored food and water.

Some sieges are unintentional, and produce no victors. As our unstable climate reduces rainfall, runoff, and storage capacity while accelerating evaporation, water scarcity worsens throughout eighty countries to pinch down on 40 percent of the human population. Rich nations have belatedly come to grips with their greenhouse gas beast and are now taking steps to restrain its impacts. But even under the best of circumstances, even if all emissions were capped immediately and no further carbon was thrown up into the sky, the world will keep warming. Populations will keep growing, snowpack and ice caps will keep melting, tropical belts of aridity will keep expanding, converting forest into savanna, and savanna into semi-arid deserts that look like the Kalahari. To paraphrase Pogo, in our conquest of nature civilization has managed to besiege itself.[15]

Yet siege remains unpredictable because the cutoff of water, while devastating, was not always or immediately fatal. At its peak in the first century A.D., Rome—which knew all about harnessing water as a political instrument—encountered in Judaea the Great Jewish Revolt and besieged a thirteen-hundred-foot-high plateau overlooking the Dead Sea. Masada became the last retreat of a tribe of indigenous minorities, seen by Rome as an "extremist splinter group" of Zealots. Unlike the occupants of other besieged Jewish fortresses, the people of Masada chose not to counterattack the Romans. With enough access to natural water, they had no reason to. Time was on their side as the siege lasted for years, devouring Roman manpower, money, energy, and reputation. It apparently ended only after the indigenous people decided to choose self-sacrifice over submission.[16]

Military historians stress the extreme risk and volatility of sieges. Nothing is certain. Depending on each side's ability to deny pain, on its tolerance for sacrifice, on its self-discipline, and on its ability to adapt to extreme conditions, the very forces designed to weaken the besieged can just as easily break down the attacker.

PART II

Adaptation

CHAPTER 6

The Rule of Water

Ancient scriptures and modern westerns warn us of risks inherent to the desert. The Latin root of *desert* was originally equated with wilderness, a people-abandoned place of spiritual desolation. Out there in the remote desert wilds lay a lawless existence, a physical and amoral danger zone lurking beyond the boundaries of decency, a land ruled by turmoil and existential doubt. In the desert Moses's people wandered in anguished confusion, and even Jesus wavered. Out there, "chaos was the law of nature," affirmed Henry James. "Order was the dream of man."

From ancient Rome to modern nation-states, civilization imposed order on wild nature, lest nature impose its chaos on it. As Charles Darwin prepared his *On the Origin of Species,* what Victorians feared most was that the amoral blade of natural selection might lay waste to the existing benevolent theological order.

> *Who trusted God was love indeed*
> *And love Creation's final law*
> *Tho' Nature, red in tooth and claw*
> *With ravine, shriek'd against his creed*[1]

Nowhere did that law of tooth and claw appear to menace more than in Africa. Old maps filled blank spots with fierce beasts and savage humans, and even in the twentieth century, *National Geographic* and *Animal Planet* documentaries filled suburban living rooms with the Dark Continent's rat-eating plants, snake-eating frogs, bird-eating spiders, man-eating lions, and monkey-eating humans. News reports suggested that Africa's indigenous tribes lacked any overriding laws to check their passions, resulting in unbridled lusts, child soldiers, superstitious cannibalism, and genocide. Yet a closer look at the forces guiding Africa's desert wilderness revealed less random chaos than structured order, less anarchy than a natural, and even rather beautiful, dictatorship.

Indeed, as science revealed the intricate mechanisms of natural selection, hardened Darwinian atheists suspected an almost religious Common Destiny at work on Earth.[2] The elemental catalyst behind this higher purpose was not oxygen, sunlight, fire, or soil; none of these were essential to maintain all life forms. The force that governed existence in dry lands was the relentless quest for water.

Organisms evolved through the desert's unforgiving but impartial justice. Moisture-capturing and -retaining species won eternity: reproduction. Failures met the end of birth: extinction. Out of this arid dictatorship emerged a profusion of intricate adaptation by all plant and animal species, including *Homo sapiens.*

Adaptation didn't appear spontaneously, out of nowhere. It evolved literally at a glacial pace, in response to incrementally drier conditions. As ice ages advanced and receded from polar regions, climate gradually transformed the Kalahari.

The Bushmen's reserve wasn't always so dry. Satellite images reveal the shoreline of what was once a vast, ancient body of freshwater, when Botswana's Kalahari heartland had once been submerged, flooded by as many as three great rivers. Two million years ago it held a 150-foot-deep "superlake," with its own weather system, that may have resembled North America's own, now gradually shrinking, Great Lakes.

What happened to all that water? First, earthquakes. As an extension of the Rift Valley, Botswana's northern edge has always been tectonically active. A million years ago geological plates collided and gradually lifted up southern Africa's central plateau. The effect diverted the rivers off to the Zambezi, the Limpopo, the Orange; the Okavango backed up onto itself and became a stagnant swamp.[3]

Heat followed the plate tectonics, as a few degrees' rise in temperature lifted off 96 percent of the region's surface water. Elsewhere in tropical Africa, a half-dozen millennia of evaporation converted a 1,000-foot-deep lake into what is today a bone-dry crater, while the 154,000-square-mile Lake Chad—a body of water larger than the Caspian Sea—contracted exponentially to less than 200 square miles today. Millions farm its receding edges, migrating inward into the vortex as mudflats turn to dust.[4]

In the Kalahari, the earliest Bushmen ancestors might have encountered a body of water one hundred feet deep, but by ten thousand years ago would have found nothing but the giant Makgadikgadi Pans.[5] These flat, seasonal pans remain phenomenal. For a few months, after brief thundershowers, short grass in the pans attracts innumerable zebra and blue wildebeest, but most of the year they spread out as hard and flat as concrete.

As lakes evaporated into pans, their flowing sources also shrank. Roiling rivers diminished to feeder streams, streams became dry beds, and beds eventually sank down into depressed fossil creeks. Bushmen, who know these fossil creeks intimately, explain how their meanders were carved out as one of their gods, Ga'mama, dragged his snake-bit leg across the sands in search of waters to drink.[6]

In a less mythical way, thirst also sculpted the Kalahari's living inhabitants. "Most plants and animals survive in the desert," explained Africa's foremost desert ecologists, "because they do not live in the desert."[7] Natural selection equipped organisms with the genetic tools and inherited wisdom that enable them to tolerate, avoid, or escape the most extreme and unforgiving conditions of heat and water scarcity in the desert. Plants and animals retreat to become dormant and inactive during the worst hours, days, seasons, or years in the desert. Alternatively they migrate, trekking with the seasonal and annual variations over long distances. Under the arid order these physical and behavioral traits coevolved, first in plants, insects, and increasingly complex vertebrates; then in our own species.

Drought's dictatorship shapes Kalahari vegetation. The camelthorn acacia sinks taproots to extract every precious drop from pores between deep sand.[8] The baobab's spongy bark is saturated with water to endure protracted droughts. The Kalahari "rain tree" attracts insects that scatter down moisture on cloudless days.[9]

Grasses and shrubs chase down water in slow motion. Vines, creepers, tuberous species lie dormant for years waiting to reach out and pounce on any microscopic drops that meander past. Lichen snag water vapor from the air; ground-hugging leaves pin down night moisture; roots hunted seep lines for damp sand deep beneath the surface. All plants cling fiercely to whatever water they catch. Some lie on wax, compete for precious shade, follow the solar arc to reduce exposure, or stop breathing until nighttime. To reduce moisture loss, trees shrink leaf size, fold leaves during the day, or drop them during the hot dry seasons. Certain Kalahari leaf patterns even funnel precious drops of rain inward and down to the bulb or root structure beneath the ground.[10]

The same ruthless order governs desert insects. They absorb moisture and then seal it in watertight, even at the expense of flight. Spiders drink moisture collected on webs, while beetles dig trenches to channel water into canals, or collect dew on their skin, then direct it toward their mouth to guzzle the equivalent of a human chugging seven gallons in one session.[11]

More complex adaptations come with backbones, as toads, tortoises, snakes, birds, and mammals seek water from what they eat, having also evolved ways to retain that water.[12] Kalahari dryness has elongated ground squirrel tails for shade and dictated the colors, patterns, and thickness of zebra and antelope hides to reflect sunlight and hold moisture beneath the skin. The most durable desert antelope prefer plants with short life spans, chosen for water content rather than nutritive value; they also feed mostly at night when even dead grass and leaves grow swollen and heavy with absorbed atmospheric moisture. The survivors lick morning dew; dig up succulents, melons, wild cucumber, and underground plant structures; and their metabolism concentrates urine and excretes dry feces. The largest mammals instinctively breathe slower, seek shade, or minimize exposure by facing the sun. They are armed with complex nasal passages that cool cranial arteries like radiators. Their brains unconsciously let body temperature rise up to thirteen degrees during the day, then cool during the night, rather than sweat the precious two to three gallons of water that would have been sacrificed through evaporative cooling. Constant aridity drives predators to kill as much for thirst as hunger, and before digesting protein in antelope meat, the lion, leopard, and hyena often lap up the blood, moisture, and fluids of intestines.[13]

In short, thirst imposes a strange order that compels plants to perform self-mutilation of leaves, drives "pure" insectivores like yellow mongoose to eat succulents and "pure" carnivores like cheetah to devour tsama melons. Atop the Kalahari water pyramid stand the Bushmen, carefully observing the surrounding organisms, mimicking their desert-savvy ways.

Natural selection had endowed Qoroxloo and her ancestors with very few built-in physical adaptations. True, her size, pigment, and hair help offset the risks of extreme exposure. Yet beyond that, Bushmen are inherently no better equipped than any other humans. Instead, their songs, legends, and experience provide the only real evolutionary advantages that they have inherited, an unwritten code that enabled them to evolve in the Kalahari and coexist in peace, despite the slow but inexorable vanishing of the waters.

Their existence further refuted Henry James's dictum about nature's chaos and human order. Reckless people may ignore laws, but habitat does not. As the anthropologist George Silberbauer observed decades ago, "It is plain to see that [the central Kalahari] is an ordered ecosystem. Were it not ordered, the Bushmen would perish, for they, like the rest of mankind, have not the capacity to adapt to chaos."[14]

The discipline imposed by the rule of water was not always evident or easily detected, since it was marked by extreme inactivity. During the worst heat of the day Qoroxloo mostly busied herself by doing next to nothing. Conserving her energy like the desert antelope, she sat in the shade, caught slight breezes, and inched her body beneath tree trunks and canopies, shrinking exposure against the solar arc like a human sundial. Other than that, she minimized movement, even any unnecessary gestures. She ate little, kept silent, and breathed through her nose to reduce the moisture loss sacrificed through the surface of an exposed open mouth.

In ways that resembled the foot-lifting, debris-seeking gecko, Qoroxloo sought to minimize heat absorption through insulation. There was nowhere hotter than the sand surface, so to reduce contact she always tried to keep something—grass, cardboard scraps, blankets, or mats—between herself and the physical Kalahari. Indeed, one reason she and other Bushmen welcomed tourists' cast-off T-shirts, pants, shoes, scarves, and baseball caps had little to do with fashion or status. Rather, this light clothing protected their skin from exposure to drying winds and direct rays; it absorbed sweat and held it against skin for maximal cooling effects. All this dramatically lowered her daily water requirement and thus reduced her need to go off in search of more.

And in emulation of species like the scrub hare, Qoroxloo became dormant, crepuscular, at times even subterranean. Keeping cool, she could strategically think about how to meet needs with minimal effort. To maximize results, Bushmen conserve body water by focusing mental activities at dawn and dusk. Under extremely hot conditions, Qoroxloo would try to lower body temperature by burying herself in a patch of sand, soaked with urine, to cool down.[15]

Certain bands of Kalahari Bushmen harvest a dryland species of succulent that outside of Botswana has become synonymous these days with Internet spam. This new, clinically proven wonder drug, derived through a multinational pharmaceutical company, has become a lucrative consumer good by promising an increasingly fat world that it would help buyers "lose weight . . . and keep it off."[16] Qoroxloo didn't have an obesity problem. But *Hoodia gordonii* mimicks the effect that glucose triggers on nerve cells; its chemicals trick the brain into thinking and feeling like they are sated, thus curbing appetites. How strangely appropriate that this quintessentially First World drug first had been discovered, tested, tried, and "marketed" in that outdoor laboratory, the Kalahari, by Bushmen who ate *Hoodia* before and during long hunting and gathering expeditions far from base camp, not knowing when they might next find a reliable access to water, in order to suppress their thirst.

Qoroxloo's metabolism was physically equal to that of any other human, but her behavior and adaptations were superior. Passive resistance and apparent inactivity are often mistaken for sloth or surrender. But Aboriginal people in arid Australia, Africa, and America are about as "lazy" as a snake or lion. They habitually reduce energy expenditure in a deliberate way, to conserve precious water for only essential demands. The ability to do nothing should never be underestimated; most humans grow restless and agitated in the heat. Inside the Kalahari reserve, Qoroxloo's calculated restraint revealed the oldest, most highly energy-efficient, and, from an evolutionary standpoint, most effective form of living within the rule of water.

OF COURSE, IT was not the only variety. The Kalahari's timeless dry autocracy disciplined Qoroxloo in her formative years at the same time in history that, far from the reserve, Western civilization confronted its own ruthless and expansionist dictators. On June 4, 1940, as Hitler's tanks rolled over Europe and England found itself under siege, Winston Churchill rallied the House of Commons in a rousing speech. He urged America to enter the fray against the Nazis. He ruled out surrender. He called for struggle and sacrifice and the discipline of a stiff upper lip. He vowed: "We shall fight in France, we shall fight on the seas and oceans, we shall fight . . . on the beaches, we shall fight on the landing grounds, we shall fight in the fields and in the streets, we shall fight in the hills." Yet the famous rhetoric overlooked what turned out to be a most pivotal theater of combat. During the first and most decisive years of World War II, battalions under Major General George S. Patton, British field marshal Bernard L. Montgomery, and German field marshal Erwin Rommel clashed across North Africa. It turned out that we shall fight, foremost, in the desert.

As armies ventured into the desert, that dry hot landscape presented an evolutionary crucible. Those who could best adapt would proceed; those who failed to adjust would not. History caught all sides woefully ill-prepared for such arid terrain. The United States scrambled to ready itself as its army hit the books, suddenly fascinated by the relationship between man, water, heat, and sand.

Almost overnight an entire science, *dipsology*, emerged as the study of thirst. Where exactly did thirst come from? What triggered it? How far could we push it? Could it be controlled? Existing research fell woefully short. Until then it had been rather difficult to round up volunteers eager to dehydrate themselves to death in the name of science. But some indirect and often chilling accounts offered America a starting point. Countless observers described in detail the progressive suffering of dehydrated and nauseous Africans locked beneath sweltering decks and argued there was "nothing which slaves during the middle passage suffer from so much as want of water."[17] By the nineteenth century, hypotheses defined *thirst* as not so much having a dry mouth, but rather "a general sensation arising from lack of water in the blood." Later accounts detailed how intravenous water and salt solutions relieved thirsty dogs and cholera patients, and by the 1930s some indicated that thirst arose from dehydration both within and between human cells.[18] Beyond these inexact reports, shockingly little was known. But after the outbreak of World War II, America's military researchers explored whether dehydration and thirst could be reduced, treated, prevented; perhaps even somehow cured.

These researchers turned California's dry badlands into an outdoor lab. A team led by E. F. Adolph sought to discern how far otherwise healthy soldiers could walk in the desert, under what kind of heat, before collapsing from lack of water. Observations were precise, measurements exacting. Water loss was calibrated either by decreased water intake or, less pleasantly but more realistically, by increased sweating, vomiting, or diarrhea. In order to bring uniformity to relative conditions of general thirst, lethargy, fever, muscular weakness, and anorexia, the government-funded researchers developed scientific charts that put soldiers' water loss in context, and then compared results. They weighed soldiers like groceries to calibrate metabolic stages of dehydration. A mere 2 percent drop in body water caused the brain to have difficulty with basic math, made it hard to read an instruction manual, and weakened short-term memory recall; with 5 percent of the body water lost, men grew "fatigued and spiritless." Beyond that point, eyes retreated into sockets, and delirium meant the poor enlisted wretch was knocking on death's door.[19]

The army determined to correct, compensate, or overcome such weakness, based on the following reasoning. A soldier's lung tissues expanded to live and fight at high altitudes or swim underwater; his muscles hardened from "humping it" through mud along great distances with a heavy pack; and his psyche adjusted to hours of boredom punctuated by moments of sheer terror. So it seemed logical that his kidneys, sweat glands, bloodstream, and lungs could and would also adapt, and learn to metabolize life more efficiently, when restricted to less and less and less water.

The army embarked on a rigorous training regime known as "water discipline." As drill sergeants sculpted mush into fighting men, they progressively rationed the training soldiers' supplies, denying thirst and depriving them of canteens, day after day. The idea was to toughen up the enlisted men into leathery, drought-proof warriors.

Even the fittest began dropping like flies. They suffered from heatstroke and dehydration. America's armed forces endured "thirst casualties" before any of them could reach the African front.

Mercifully, the army's "discipline" was halted. But thirst research continued, based on field observations. As it better understood the relationship between soldiers and water, army priorities reversed. Rather than adapt soldiers' thirst to fit the demands of the army, the young Pentagon began to adapt its own collective goals, targets, tactics, logistics, and requirements to meet the individual's demand for water.

If water was scarce, adaptation meant soldiers must carry an increased amount of water with them, in order to be able to march a certain distance under certain temperatures without expiring. Under tight conditions, adaptation for soldiers meant staying out of the sun and rolling down sleeves to retain sweat. The disciplined soldier was trained rigorously in how to find, extract, and access freshwater from rocks, sand, animals, and vegetation under the harshest conditions, even in the desert. Right down to mundane uniforms, Schwarzkopf boots, and MRE food menus, adaptation to water requirements shaped—and continues to shape—strategic decisions over where, when, and how we shall fight in the desert.

Water discipline slowed and minimized water loss, and would help individuals adapt to dry heat, whether they were under siege in the Kalahari like Qoroxloo, or deployed to Iraq in the summer, or caught in Las Vegas or Phoenix during a power failure that disrupted air conditioning and water pumps. But discipline alone couldn't stop dehydration. Like all species, humans eventually have to replenish supplies of moisture, electrolytes, and nutrients, and that required mobilizing efficiently.

CHAPTER 7

Dispersal

AS THE HEAVY GOVERNMENT TRUCKS drove off loaded with their precious human cargo, the Bushmen who had succumbed to fear and thirst gazed back at those who chose to stay. Eventually the dust, stirred up by the convoy, settled back down in the quiet reserve.

Qoroxloo Duxee looked around. Circles remained where round thatched huts had once stood. Gone were some of the big game hunters, and many women with whom she used to gather wild food. She talked with the others about whether the government would come back, or if President Mogae had done his worst and might finally leave them alone. The Bushmen decided to proceed on the assumption that the outside world would soon forget them and leave them on their own. So each looked to one another, and to their resources, and began to plan. Ironically, the government's clean and brutal forced departure had brought one unintended silver lining: It relieved ecological pressure on those who remained.

The average density of hunter-gatherer groups worldwide is one person per square mile.[1] In the Kalahari it was one per ten-to-twenty square miles.[2] Deprived of supplemental government supplies, a more densely packed human settlement could not long endure here. Large numbers of concentrated families would quickly have devoured and drunk most naturally occurring food and water within a short distance of home.[3] As it was, the remaining Bushmen—to seek and secure a sufficient amount of food and water—would be forced to disperse.

Dispersal was one of the oldest ways Qoroxloo's people adapted to dryness.[4] The act sounded simpler and less consequential than it really was. Hourly, daily, seasonally, and even interannually, Bushmen bands would break up into smaller units and thin out in different directions across the savanna, and in so doing reduced pressure on natural resources in any one place at any one time.[5] That physical reduction in turn decreased social friction, political tensions, and the spread of illness.[6] As an optional tactic for collective risk reduction, evolving over millennia in the Kalahari, dispersal worked.[7]

Bushmen may be casually portrayed as *nomadic*, a word that typically conjures a one-dimensional image of people wandering randomly in a clueless horde, scavenging whatever they might stumble across. The truth is emphatically the opposite. Dispersal is more strategic and exacting and focused than rock climbing. Also like that activity, dispersal requires an intimate knowledge of place, of carrying capacity, and absolute trust in the strengths and weaknesses of your companions.

Qoroxloo's band spoke often and openly of individuals' various skills, weaknesses, and behavioral temperament. Those who moved might cover a territory of four hundred square miles, taking with them an intricate knowledge of the distribution of specific food plants and animals. Men and women grasped which landscapes held priority assets at various times of the year, and how many people each area could support, for how long. Before dispersing on even a day's expedition, Qoroxloo would mock various recommendations. But her seemingly casual banter quickly teased out: which recent foraging expedition was most successful, who could best lead a smaller group, or how certain combinations might get along under the pressure of heat stress. Because each hot day required consuming precious energy and water, Bushmen did not often stir without giving that effort a great deal of deliberative calculation.

Humans evolved to drought conditions by dispersing into small, face-to-face groups, adjusting their size to the relative aridity of the land. From North America to Australia to southern Africa, a few hundred people might seasonally disperse into bands of fifteen to fifty people during the dry season.[8] On a daily basis these smaller bands dispersed their hunting and gathering pressure to even smaller, tighter groups, so that in arid lands, each hunter-gatherer had several square miles all to him- or herself. As resources grew scarcer and tensions rose during dry seasons, Qoroxloo's band would fragment and disband to other satellite parts of the central Kalahari for hours, days, weeks, or even months at a time, then coalesce when food supplies became more abundant. In any case the strategic decision to disperse would be dictated only by ecological necessity and planned well in advance. No single person enforced or led these groups. Efficient movement had to be deliberate, voluntary, and internally consensual, not driven in haste by external or reactionary forces.[9]

OUTSIDE THE RESERVE unfolded a very different story. Today tens of millions of industrialized citizens have inverted this evolutionary instinct, migrating away from dispersed homelands, only to roost in harsh arid landscapes.

Affluent tribes migrate by choice. Hordes of short-term nomads from Europe fly south to overwinter in Namibia, Botswana, and South Africa, congregating intensively in dry hot spots and single-handedly depleting water precisely when supplies are most scarce. Longer term, North America has drained much of its dispersed and rural populations from its wet and cool states to swell rapidly drying cities—ninety-two of America's one hundred fastest-growing counties are in the heatstroke-prone South, Southwest, and West—where people continue to arrive even as water supplies evaporate.[10]

The poorest tribes are forced to move. Decades ago, when drought drove millions of Dust Bowl "Okie" farmers from the southern Great Plains, most drifted not northeast toward the rain, but rather into the even drier Southern California desert. Today four thousand Chinese villages face total abandonment due to the expanding Gobi Desert,[11] while protracted dry spells and failed wells herd farmers into already water-stressed Mumbai, Beijing, Sanaa, Sydney, and Mexico City, joining the ranks of the world's estimated 25 million "drought refugees."[12]

Drought migrants have congregated in arid megacities to find jobs, hoping to earn cash to purchase water they couldn't access and food that wouldn't grow. Yet urban billions soon discovered what dissident Bushmen already knew: As droughts worsened under global climate change, food and water deliveries grew more and more strained, more expensive, more unreliable, more curtailed, and more often hoarded. The recent global end of cheap food and water helps disprove a dangerous and deadly urban myth. Despite the buzz of activity, migrants to cities in developing countries rarely found healthier, better conditions than those who stayed in widely dispersed landscapes. To the contrary, the world's one billion slum dwellers experienced more hunger and disease, and found less work, food, and water, than did their rural counterparts.[13]

Botswana officials ridiculed the semi-nomadic dispersal of Bushmen in their calculated "scavenging for food" and "chasing after water." But elsewhere that adaptive survival strategy didn't seem so backward. Critics began to warn that those two thousand people who were flocking each week to cities like Phoenix were soon finding them uninhabitable due to the rising heat and vanishing waters.[14] On the eve of the 2008 Olympics, a heated debate surfaced over whether and how Beijing's population should disperse somewhere wet enough to support a capital city of 20 million people. Rather than mobilize ever more water from the rainy south to the densely populated dry north of the country, China's leaders began to consider whether to encourage people to migrate closer to rivers.[15]

Admittedly, shifting entire populations to follow rainfall appeared radical and extreme, possible only under a command-and-control regime like China's. But to free, autonomous people like Qoroxloo such dispersal seemed merely obvious.

By late afternoon after the convoy departed, the hungry and thirsty Bushmen gazed out at the Kalahari they knew well. Working together, Bushmen shared memories, information, and experience to recall what available food and water they might forage. After consulting the elders, the remaining females made plans and discussed where next to forage in the cool hours of the following morning as the men sharpened their knives and spears.

The government had left them to provide entirely for themselves. *So be it*, they thought. *Since the water will no longer come to us, we will use old and new tactics to seek it in the places where only we know it resides.*

CHAPTER 8

Forage or Farm?

SINCE THE DAWN OF *HOMO SAPIENS* IN ARID Africa, nine tenths of our evolution has unfolded as foragers. Only relatively recently did our species embark on agriculture, and recent events suggest certain limits to that extraordinary experiment. Exponential population growth has combined with unprecedented climate change until half the planet's terrestrial surface can now be classified as drylands—arid landscapes inhabited by a third of humankind.[1] After decades of growth, arid nations as diverse as India, Mexico, China, Egypt, Yemen, Israel, Australia, and Botswana lack sufficient water to efficiently grow enough food to feed themselves.[2] To make up the difference, hungry billions have depended on foreign aid or trade, but even that pressure valve has begun to close. By the early twenty-first century, subsistence and irrigation farmers, long dependent on reliable rains or cheap water supplies, rarely still knew exactly what to do or where to turn for food.[3]

Qoroxloo did.

On mornings cool enough to walk yet warm enough so lions would rest in the shade, she strode out into the Kalahari armed only with her digging stick, her leather bag, and her female friends. Part of her goal was of course to gather nutritious wild groceries, yet each search involved a far more complex undertaking than that.[4] Qoroxloo observed emerging details for immediate use or to be filed for later. She noted an item, discussed it with others as they walked, and mentioned it again after returning home at the end of the day: a ripening fruit here; day-old giraffe tracks there; an ostrich moving into the pans. Pivoting her head while striding at a rapid yet graceful and unhurried pace, Qoroxloo scanned the horizon, the sky above, and back down again, forever seeking latent signs that suggested the presence of moisture.

Carnivorous hunters had to think not like the herbivores that drank the arteries of plants; Qoroxloo had to think like the plants that sucked water from the moist capillaries in infinite sands. In this all-consuming search for moisture even bugs took on significance. Mopane worms, locusts, and other slow-moving grubs offered succulent, juicy food, while flying insects served as beacons to damp places. Water vapor sent a subtle signal into the air that insects sniffed out and hunted down: A congregation of gnats suggested local humidity; butterflies alighted on damp earth; bees flew in a direct line toward any source of standing water, and she followed their cue.

Yet Kalahari adaptation brought an inherent tension. Drought-resistant plants sought to capture and retain water without advertising themselves as moist targets for consumption by more complex species. By hiding and hoarding their wet bulk deep underground, plants revealed no more than a thin stalk and waited for just the right moment to convert energy and moisture. Many animals overlooked that stalk, but not Qoroxloo. She especially tried to detect and investigate closely with others the plants that Bushmen knew as rain-swelled "storage organs," their substitute for water during the ten months when Kalahari pans were dry.[5] As she educated the next generation about how and why "some plants were for eating and other plants you can drink," her lessons boiled down to seven habits of highly successful foragers.

First, harvest every raindrop embedded in the flesh of plants. Qoroxloo's task was challenging when even fleeting showers sank quickly out of reach and less than 8 percent of Kalahari sands absorbed water and remained damp. The sun dried and the wind blew most leaf debris away before it could decompose into rich organic humus. Still, she knew to seek plants and tubers, like marama bean, which gorged on passing cloudbursts until some specimens hoarded up to two gallons in their bladder cells and stems. All in all, from this indifferent dry landscape foragers like Qoroxloo harvested as much as 90 percent of the band's moisture, contributing two thirds to three quarters of the band's food, the bulk of its caloric intake, and more nutrients and oils than obtainable from the meat men could hunt.[6] It was quite a crop for such few precious drops.

Next, defer decisions to the authority of the local expert who knew where, how, and when to harvest. Efficiency required democratic collaboration because, during the hottest months of the year Qoroxloo faced a precarious battle to gather more moisture and nutrients from what she dug than she burned off in the process of digging. Timing was everything. Even after reaching a swollen underground tuber, Qoroxloo might choose not to harvest it immediately, but rather break off a small sample. If nourishing and tasty, into her bag it went; if bitter and shriveled, she replaced the sand to return at a ripe date.[7] Her extensively mapped knowledge of stored provisions could be exchanged informally with other foragers; in this way Bushmen minimized waste, avoiding the harvest of too much, too soon, or too late. Qoroxloo did not plunder the Kalahari to depletion, but tactically cultivated a wild, permaculture garden to which she might return.

Third, diversify the diet as much as possible. This wasn't a matter of taste, but survival. By utilizing the widest spectrum of what was palatable—some 70 to 150 species of roots, rhizomes, tubers, melons, berries, and fruit; plus 40 animal species—Bushmen broadened the foundation of their vegetable food security. True, some items on Qoroxloo's menu were admittedly tasteless but still highly nutritious, or laden with moisture. The vast and complex fine art of plant identification and use, and avoidance of poisonous food, further confirmed that Qoroxloo's band could not have recently or suddenly "retreated into the desert," as some claimed, but had coevolved with Kalahari species over millennia.

Fourth, eat locally and seasonally. Distance opened more opportunities but consumed precious energy burned up in transport to and from. Central stockpiling attracted pests and raised the odds of rotting food. Throughout the dry months Qoroxloo and her band moved from one patch of water-embedded plants to another, and left as much food as possible stored underground in decentralized caches. During this time Bushmen ate eleven pounds a day to get the liquid they required.[8] Whenever possible Qoroxloo and her fellow foragers ate as much ripe and moist food as they could at the far end of their journey rather than carry it miles back home. In short, Bushmen could satisfy their daily minimum water requirement by seeking the shortest distance from the earth to the hand to the mouth.

Fifth, scale back efforts to the least disruptive technologies. It's hard to imagine a more functional, timeless, and durable tool than Qoroxloo's digging stick.[9] Digging punished haste and sudden exertion but rewarded the pacing of energy as Qoroxloo chopped and picked, then scooped out the loose sand. Tock, tock. Scoop. Tock, tock. Scoop. She bore down in the blazing heat, hunched over, pausing every few inches to catch her breath, tracing a vine down to the damper parts of the sand like an archaeologist gently brushing away sand in a search for answers. Near the root she slowed down, lest the stick damage the plant. The second technology with which Bushmen manipulated landscapes was fire. Their small-scale burns created a rich mosaic that encouraged new growth, attracted prey, and reduced water lost through old dense vegetation.

Sixth, innovate as much as the habitat allows. Bushmen were not locked in the past, wedded to tradition. When they noticed how tsama melon seeds took root after sudden rains, some in Qoroxloo's band experimented with these and other emergency fallback sources of food. Doing so further blurred the textbook line between farming and foraging, and government officials cited the existence of a few small rain-fed melon or millet gardens or goats as clear proof that dissident Bushmen had been corrupted down the slippery slope of agriculture and pastoralism like the rest of us, and so must be forced to move out to join the rest of "civilization." What remained unclear was whether Bushmen were wrong, or the textbook categories were incorrect. In fact, anthropologists recently discovered that shifts from hunter-gatherer to pastoralist and farmer were seldom one-way, irreversible stages of teleological progress toward a more complex and perfect existence. Rather, *Homo sapiens* had always slipped back and forth across a variety of survival strategies, mixing and matching techniques of food provision to fit the conditions of the landscape.[10] Throughout much of our evolution, the determining forces of our human diet rested largely on rainfall and access to water.[11]

Finally, subordinate taste to survival. In the quest to gather moisture along with food, two thick-walled vegetable canteens stood out in particular: Gemsbok cucumber held 70 percent water, on average more than a pint, and while bitter they could be chewed up, moisture extracted, and the fiber spat out. Better still were the tsama melons, which began to swell during the rainy season and reached their peak during the dry winter months. Their round shape minimized exposure through efficient surface area, and the average one-pound melon consisted of 90 percent water; eighteen melons yielded twelve quarts. Since these were rarely boiled, they captured the full richness of a diet high in vitamins and nutrients. Qoroxloo split them open with a few thrusts of her digging stick, stirred up the pulpy contents into a slushlike watery mash, and then slurped the contents out by hand. Indeed, Qoroxloo's band dutifully roasted and "ate" the less tasty, nutritious dry fare but "drank" juicier, more flavorful food with relish.[12]

As I was an outsider to the heartless Kalahari, Africans had warned me to fill my canteens and bottles: "You can endure weeks without food, but only days without water." But these insiders revealed in their advice a modern false dichotomy. To Qoroxloo, food was water. Water was food.

FOOD AND WATER use diverged roughly thirteen thousand years earlier, with the onset of farming. Some attribute the birth of agriculture to wetter times, others to the millennium-long dry, cold "snap" known as the Younger Dryas.[13] Either way, the human approach to food as a prime source of water progressively split in two.

If foragers quickly ate local plants that stored water, agriculture stored water to slowly grow dry cereal grains that could be eaten after months and bartered long distance. As long as there was enough water, the latter approach offered considerable benefits. Following this split, the world shifted from rain-fed subsistence crops to exploit arid lands, extracting more and more high-value food reserves—especially monoculture or single-crop grains—for storage, processing, trade, and shipment to cities. Egyptian pharaohs instructed slaves and feudal lords ordered serfs to plant grain in advance of the rains; industrial tools accelerated this trend; fascist and socialist and capitalist leaders alike urged dryland farming, issuing propaganda about how to "make the desert bloom" since "the rain follows the plough." When such predictions proved false, engineers stepped in to correct them with dams, dikes, and irrigation diversions across every river until today more than 70 percent of all freshwater use on Earth gets converted into food production.[14]

The mobilization of water transformed food and diets as well, often for the better. Yet only a negligible fraction of our food retains its natural moisture. And now we're not only expected to feed another 3 billion mouths in the next few decades, but twice that many mouths will hunger for the richer diets enabled by increasing affluence.[15] Wealthier humans may require less than a quart of water to survive, but we routinely "drink" three to five thousand gallons in the meat and grain of what analysts have christened the *virtual water* that gets embedded in the creation and processing of our food.[16] The term is apt. Thanks to evapotranspiration, all that water lost en route to the table is never again tasted or seen. Economic affluence further distorts this water/food dichotomy: The luxury of a soft drink containing sixteen ounces of water requires hundreds of gallons to grow the sugar, "high fructose corn syrup," or "evaporated cane juice" contained within. Consider how much virtual water it takes to grow just one healthy American breakfast. A bowl of Rice Krispies with milk: 350 gallons. Dry grounds for a cup of coffee (sans real water): 37 gallons. Eggs and toast and a side of bacon: 470 gallons. Multiply that pressure by 6 billion people, 365 mornings a year, and you get an idea of the scale of the planet's growing stress from the water footprint of irrigated agriculture.[17]

Even chronically dry nation-states such as Botswana could not resist plunging into this arid land irrigation revolution. In the name of "nation building," "job creation," and "food independence," Botswana's National Water Master Plan aimed to provide a fifth of the country's grain consumption within two decades. That meant in addition to subsidized tractors, fuel, fertilizer, herbicides, and pesticides, President Mogae's government doled out prodigious quantities of its freshwater. It pumped some 18 billion gallons from Kalahari aquifers or the anemic Limpopo River, and then sprinkled all this precious finite water out onto its increasingly hot sandscapes. It planned to quintuple its irrigation area crops such as corn, alfalfa, vegetables on four thousand acres. Most crops dried up. But not before Mogae's government reaped UN Food and Agriculture Organization development aid funds and harvested loyalty from a subsidized workforce.[18]

In fairness, Botswana was hardly alone. Saudi Arabia burned oil to desalinate seas to irrigate wheat, Egypt bled the Nile into the Sahara to grow grain pricier than it could import, and the United States poured out water on burning western deserts to grow fuel substitutes. The irrigation addiction was easy to rationalize. Given productivity gains from, say, India's Green Revolution, intensive dryland agriculture brought the sudden rush of food security and temporarily warded off famine for billions. No one could really complain, until everyone did, all at once, as supplies dried up around the world.[19]

In the United States, the Ogallala Aquifer bottomed out across America's breadbasket Midwest even as Colorado River reservoirs stored half of what they could a century ago. Entire irrigated regions of China, India, and Australia withered as the Yellow, Indus, and Murray-Darling rivers, respectively, stopped flowing. Lake Chad and the Aral Sea dried up into shallow ponds, and climate change altered the rainfall, runoff, and evaporation patterns for a third of the world's population.[20] To make matters worse, in recent years up to a third of global water withdrawals were converted to irrigate new sources of fuel.[21] Since it takes nine thousand gallons of water to produce a gallon of biodiesel, and four thousand gallons to produce a gallon of corn ethanol, a "clean fuel" drive to the supermarket could burn up more water than would otherwise have grown a week's worth of calories. "There was a remarkable lack of careful planning in the drive to convert food into fuel," wrote Nestlé's chairman, Peter Brabeck-Letmathe. "Either it never occurred to biofuel advocates to ask about the amount of water needed for biofuel production, or they simply chose to ignore this particular inconvenient truth."[22]

The resulting food price inflation brought widespread misery on scales so deep and pervasive that it defied statistics. As a rule of thumb, middle classes cut meat, lower classes surrendered vegetables, and impoverished billions stopped feeding themselves and turned to food aid.

Some political and business leaders, knowing that in the decade ahead the world will grow hotter and add a billion more hungry inhabitants, have begun to question long-standing assumptions about dryland irrigation. Meanwhile agricultural experts trace the origins of the global food problem to its figurative and literal roots, weeding out extraneous factors to grasp one essential priority that even Thomas Malthus couldn't appreciate. In order to feed itself at its current pace, even without a third of all irrigation diverted to biofuels, the world will deplete 2.9 billion cubic miles of freshwater by 2050.[23] Food production isn't dramatically limited by land, innovation, price, know-how, support, fertilizer, or technology. It is cut by the loss of water.[24]

Upon reaching this broad verdict, the experts developed a remedy: squeeze more productivity from each raindrop; devolve water authority to informed local decision makers; diversify crops beyond risky monoculture; reduce distances between production and consumption; scale back water-intensive methods; adapt crops to local rainfall; and exploit "green water" embedded in the ground.[25] This agenda for the world looked solid on paper—and mirrored the Bushmen's own seven-point prescription—but was proving hard to swallow outside the Kalahari. Powerful organs of the body politic felt threatened and rejected it. Even supporters conceded that this water-based or hydrocentric approach to food production required a daily discipline that was more intuitive art than technical skill. It demanded the experience and training of someone whose families had been living this "new paradigm" for centuries, someone like Qoroxloo.

Qoroxloo showed women how to bag their quarry of wholesome, moisture-rich plants. But while two-dozen melons yielded more energy than a two-pound sirloin, they didn't taste half as good. That's when she helped young men hunt the moisture-embedded flesh of those Kalahari species not held in place by roots.

CHAPTER 9

Quest for Meat

SOME BUSHMEN TOLD ME THAT IF YOU LISTEN carefully you can hear the stars, and one night in the Kalahari silence it seemed possible. Women and babies were sleeping in domed thatch huts. Only the men murmured around the fire, faces glowing, voices punctuated by a soft glottal "click."

"What kind of meat was it?" I inquired, after a while.

After a few moments my translator answered. "Gemsbok. Male."

"It was delicious," I enthused. "Where did it come from?"

The fire sparked a few more times.

"I mean, who should I thank for providing . . . for hunting the gemsbok?"

The men stopped speaking and glanced up. But my translator did not decipher the question. He simply shook his head and shrugged. "They don't say. And I don't ask."

Wise policy. If Bushmen shared too much information about a kill, and paid informants later snitched to the government, the retaliation could be excruciating and, in a few cases, fatal. Once the siege began, Botswana refused to grant "special game licenses" to Kalahari Bushmen, instantly transforming legitimate hunters into criminal poachers.

Among the band's youngest poachers were my friends "Smith" Moeti and his younger brothers, Chika and Galomphete. When the siege had begun, and these three prepared to venture out into the Kalahari to hunt for the first time without adults, they faced a more dangerous rival than their ancestors ever had. Lions and hyena still sought the same antelope that humans did, and were often heard coughing and snickering on the outskirts of camp, held at bay only by the fire. But the Bushmen's deadliest new competitor walked upright, was equipped with superior technology, and was drawn to fire as a homing beacon. If Qoroxloo's grandsons escaped being gored by horns or gashed by claws and teeth, they could be arrested, imprisoned, and beaten to death by the elite armed units vigilantly trying to track them down. As young Bushmen studied the tracks left by large game, they knew that they were also being stalked by aircraft and pursued by Land Cruisers. In their own reserve, the hunters became the hunted.

No elder told the young men to break the law. Qoroxloo's grandsons assumed the risks without hesitation, after deciding that the health and integrity of the band was at stake. As they described that hunt in detail, each grandson tried to explain how the high-stakes compulsion to hunt unto a kill involved more complex rationales than mere hunger; it was a quest to satisfy a deeper kind of thirst.

As they rode bareback on horses, Qoroxloo's grandsons scouted the land ahead and sands below, and hours passed without discussion. Eventually they came across distinctive tracks and dismounted. The prints were large and round enough to be identified quickly.

Eland, said Moeti.

His brothers then discussed the tracks' timing, direction, numbers, and gait. Hours earlier, they decided, several eland had been traveling at a fast walk or slow trot. As the young men rode on, they explained, each of them tried to inhabit the senses of the antelope, to smell and see as it might have, in order to become their prey.

Nomadic, gregarious, and mobile, eland stand more than two yards from the ground, weigh a metric ton, and carry the bulk of a Hereford ox combined with the speed and grace of a gazelle, able to sail over ten-foot-high fences from a standing jump. Eland are extraordinary beasts and seem to know it. As they effortlessly leap across each other, flowing over the sand like a flash flood, they announce their vigor and prowess to the world, hoping to discourage pursuit.

This has had the reverse affect on Bushmen. The eland drove Qoroxloo's ancestors to smear or carve its image on stone throughout the subcontinent and to associate the animal with rainfall. A single bull could provide enough fat-rich meat to sustain an entire band for weeks. Yet as prey the eland is notoriously elusive and skittish: Most African antelope flee human scent; others leap from our sight or sound; eland scatter at mere rumors of our presence. They become phantoms: often heard, rarely seen. Hunters might detect their pace and presence by the eerie, telltale, mechanical "click, click, click" of the bulls' knee-joint, a signal of fighting ability and mating status.[1] But in recent decades, while hunting eland, Bushmen also listen for the telltale, mechanical "click, click" of government rifles.

Since the nineteenth century, Botswana's ruling Tswana tribes viewed eland as exclusive "royal game" for chiefs. After independence, royalty moved into the chambers of the national government, which continued to deny Bushmen rights while reserving eland exclusively for big-spending foreign trophy hunters. High-powered rifles typically passed up imperfections—a broken horn, blind eye, deformed limbs, or diseased hides—and thus weakened the gene pool by targeting only the healthiest specimens.

Bushmen do not hunt trophies. The idea of mounting stuffed heads seems ridiculous to hungry people without wood-paneled walls on which to mount them; besides, the injured and sick specimens are easiest to kill with the least risk. "Hunting is not for fun, but it is an exercise and practice in congruence with identification with the game and the environment," said Moeti of his eland chase. "It is an aspect of commitment and covenant with God in that it justifies one of his purposes of creation, to live by feeding on what is at your disposal and not keeping that to yourself but sharing with others in appreciation of Creation."

Not far from Metsiamenong, one of Qoroxloo's grandsons and I watched three giraffe emerging from behind trees. Moved by the spectacle, I quoted *Out of Africa*, recalling "the progression across the plain of the giraffe, in their queer, inimitable, vegetative gracefulness, as if it were not a herd of animals but a family of rare, long-stemmed, speckled gigantic flowers advancing."[2]

"Yes, giraffe," he acknowledged after a minute of silence. "My favorite meat."

His remark cut to the heart of which species Bushmen value most, and why. Over 10 million years of aridity, Kalahari ungulates—eland, gemsbok, kudu, red hartebeest, springbok, steenbok, duiker, and giraffe—developed survival strategies that enabled them to stay alive, even to thrive, during protracted dry spells. *Homo sapiens* evolved physical traits and, some argued, logical reasoning in Africa by stalking these same fleet-footed grazers and browsers for their protein and moisture-rich meat.[3] Desert antelope operate on varying degrees of water independence; they verge on a drought-proof existence.[4]

To be sure, each animal still very much craves water; eland consume eight gallons or 4 percent of body mass as moisture each day. But while the beasts may drink from a rare puddle of surface water, they cope when, more often, there isn't one. Eland range far and wide, seeking food with enough water embedded within, concentrating thousands of tiny, decentralized water sources from hundreds of square miles into one convenient twelve-hundred-pound reservoir of wild meat.

Because they can live indefinitely in the Kalahari, these large water-independent animals share a certain quality with human beings. Bushmen call this quality *n!ow*, a word that anthropologists can only partly translate through what it is not. Predators like lion or leopard do not have *n!ow*; nor do elephants, buffaloes, or small animals and birds. Even gods do not have it. More positively, *n!ow* connects those animals who possess it with the cosmos, and with weather.[5] *N!ow* resides at the nape of the neck, that vulnerable spot where skin prickles in horror stories. When a woman gives birth and her water falls on the ground, or when a hunter kills a water-independent antelope and its blood falls on ground, *n!ow* can bring about change, as in the climate or weather, or like a rainbow after a storm.[6] Far from being a destructive end to existence, Qoroxloo's grandsons believed hunting was thus creative: Beyond an eland death Moeti's aim was also, simultaneously, to spill blood onto the sand and thus trigger the *n!ow* to bring about necessary renewal. He ended life, Moeti explained, in order to release it.

Some Bushmen attribute their high-risk drive to hunt as *instinct*, a word loaded with ancient currents. For more than a million years, whoever could eat meat, did eat meat, and rewards in stature followed. Meat protein gave primates and other omnivores certain physiological and developmental advantages that vegetarian herbivores lacked. It broadened dietary possibilities, ecological niches, expanding opportunities to be fruitful and multiply and establish themselves in new territories. Above all, meat protein raised the odds of making babies. Qoroxloo's grandsons were expected to hunt by the time her granddaughters could reproduce; a "first kill" ceremony was balanced by rites of menarche.[7] Because imperatives to seek, hunt, and secure wild meat are so strong and so ancient, antelope help transform young Bushmen in what David Lewis-Williams calls the *animaux de passage*.[8] "When I first hunted," said Roy Sesana, recalling his first kill of what became his totem, "I was not allowed to eat. Pieces of the steenbok were burnt with some roots and spread on my body."

Reproduction is a powerful motivating force to hunt, and hunt well. Nonhunters never starve; they share in meat others bring back, although rarely the choicest portions. But a male Bushman who fails to kill a large animal will never earn a wife. Anthropologists have concluded that this is an unwritten rule of human evolution: "Woman like meat."[9] Without the ability to provide it, his family bloodline will hit a dead end. Conversely, Moeti described how his "first feeling of killing is a sense of maturity and achievement of adulthood."

To arrive at that moment, Qoroxloo's grandsons trained with the bow and grub-poisoned arrow and used it where appropriate. Another technique they employed was to ram long hooked poles through tunnels to snag springhares. From childhood through old age, men and women carefully bend branches, weave cords from plants, and set snares for duiker, kori bustard, and other small game animals. And whether they do so on foot or mounted, Kalahari Bushmen practice the fundamentals of what is potentially humankind's oldest technique: the chasing hunt. "You track the antelope," said Roy Sesana. "He knows you are there; he knows he has to give you his strength. But he runs and you have to run. As you run, you become like him. It can last hours and exhaust you both." Pursued eland can travel nineteen miles per hour but tire quickly. The trick is to maintain constant flight and never let it drink or eat moisture from plants. In this way the source of the desert antelope's efficiency makes it vulnerable; by keeping it incessantly in motion for several hours over 120 degree sands and uneven terrain, sometimes sprinting, Bushmen can exhaust the antelope's delicate metabolic water balance. They have learned to turn their prey's thirst against it. This extraordinary feat was recently caught on film, showing Karoha, a Bushman hunter, in hot pursuit until "the blood of the animal boils before his own."[10]

After several hours' chase Moeti came upon the eland, surrounded by dogs, held at bay. He dismounted, holding his spear, and approached. The eland's legs had stiffened from lactic acid building in its muscles. Clicking knee joints fell silent. Its temperature could rise no further. It had lost too much moisture as it fled. Predator and prey watched each other.

Even when tired and surrounded, Kalahari antelope will fight hard against what Moeti called the hunter's "will to kill." This endgame can last half an hour as the big antelope ducks, stabs over its shoulder, and sweeps its horns along the ground to keep its antagonists at bay.

Moeti lifted back his spear and stepped closer to the giant eyelashes. He took aim at the chest, for "the heart is the central nucleus where life is emanated." If accurate, one thrust would be enough. But "killing an animal is a very tentative and fragile exercise," he said, "because when you miss it means your purpose of killing is not approved or recommended by the spirit of the ancient praying mantis."

What follows is rarely pretty. All animals, including humans, survive in the Kalahari by eating and drinking, thrust together by arid forces beyond their control. This eland died to give life. Moeti chanted a few traditional words meant to declare the meat edible and bid the animal's soul a good journey.

He did not delay in field dressing, for flesh can sour fast in the heat. In line with the customs of his ancestors, Moeti flayed away the wet hide, sliced open lines along the skin following the structural build of the animal, keeping it intact and free of sand. He cut through the carcass from the neck to the anus, and set aside the glands, entrails, sacs, and soft organs to reward the dogs. Moeti saved the blood that flowed into the cavity to be roasted later. The ruminant stomach of the eland held up to one hundred quarts of fluid. He squeezed out this liquid rumen and filtered it through leaves into empty plastic containers to be later boiled in cooking and drunk. Then Moeti severed the neck, split the rib cage, separated the fore- and hindquarters, and finally cut out and held the eland's warm, slippery, soccer-ball-sized heart—so dense, heavy, and filled with moisture.

Now came the most dangerous and vulnerable hours. If government patrols closed in during a hunt, Bushmen hid their weapons, concealed spears beneath a half inch of sand, or slid bows and arrows into nearby branches. After a hunt Bushmen distributed meat so widely it vanished quickly without a trace. But as they cut through flesh and sent a cloying smell up into a sky where vultures wheeled in an ancient equivalent to a neon sign—fresh bloody meat below!—the three poachers could be caught red-handed.

Even if Botswana didn't snag Bushmen in the act, potential suspects were treated as, literally, fair game. Police officers and wildlife patrols tracked and ran down Bushmen over long distances in a mock pursuit until human blood boiled, whereupon these hunters of men closed in on their prey. They reportedly tied suspects' feet in tree branches or lashed them to the bull bars and front bumpers of 4×4 vehicles. Qoroxloo's son, Tsuo Tshao, said, "[Officials] shackled my hands and ankles together before cuffing me to a Land Cruiser bull bar. They drove for a kilometer like that."[11]

Such tactics proved effective. Bushmen risked casual brutality if they did hunt or grinding hunger if they didn't. Some of those who had been evicted were reduced to eating their own, now useless, dogs. Qoroxloo's grandsons saw the broken spirits and broken bones of those who had been caught. Moeti knew the risks: up to five years' imprisonment; and eland brought especially severe punishment.

Five patrol officers allegedly beat Selelo Tshiamo, Marama Phologo, Tsuo, Letshwao, and Gaolikwe while in custody to make each suspect confess to killing an eland. No officials were punished after a second interrogation compelled isolated Bushmen to recant.[12] Selelo Tshiamo died weeks after his own interrogation. Such vindictive brutality seemed disproportionate to the crime, yet severe punishment of indigenous poachers predated the country, and violent confrontations always escalated during droughts, as beef grew scarce and pricey. "We've been beaten, tortured, and taken to court for hunting eland," said Kganne Kgadikgadi. "Why? Why can't we hunt in our own land, like we have for thousands of years?"

The most plausible answer has global implications. Hunting ensured Bushmen an independent and autonomous existence, even after the government cut off water. Below twenty-three inches of rainfall, wild ungulates outcompete domesticated livestock at every level and terrain, across the board: disease, predation, and drought.[13] Drought-proof antelope favor those who can track and eat and drink the wild game. Successful hunting challenges the centralized, hierarchical control of domesticated meat production, and thus economic clout. The more meat hunters can provide, the higher their status, the greater their reproductive opportunities. When Moeti brought down large royal game like eland, it suggested that he perhaps considered himself not just equal to other Africans or Europeans, but in some ways, superior. After all, he regularly ate more meat than working-class non-Bushmen could afford. He had more to offer women, and his marriage prospects improved. So from the perspective of authorities outside the Kalahari reserve, Bushmen hunters had to be stopped and punished, cut down to size, made as dependent as every other man who depended on a domesticated form of meat.

THE KALAHARI BRED a profusion of wildlife that Qoroxloo knew and used, but she never laid eyes on the one thirsty beast that surrounded her reserve on all sides, closing in.

The threat from this species wasn't unique to Bushmen. In the Amazon lived scattered bands of "uncontacted" hunter-gatherers, the Ayoreo, who lived in even more complete isolation. Their jungle had been a protective buffer until climate change dried up tributaries and beef prices rocketed. That's when Paraguayan ranchers took over the Ayoreo's traditional drinking source, diverted it for cattle, and cut off access to the springs. Outgunned and out of water, on March 3, 2004, seventeen besieged and excruciatingly thirsty Ayoreo surrendered their isolation to beg for water.[14]

That pattern of "first contact" had played out in Brazil, Australia, Indonesia, the United States, and Africa. Hunter-gatherers long coexisted and even sometimes traded with settled farmers and merchants, but eventually livestock broke the balance. Domesticated beasts require more space, more care, and more water until ranchers envelop the indigenous tribes and seal their fate. Anthropologists believe hundreds of autonomous tribes still exist worldwide, fighting mostly in vain to stave off cultural extinction as herders close in on their scarce water, just as they had since the dawn of pastoralism.

Back then, the now-extinct wild aurochs was reportedly temperamental, ferocious, had long, impaling horns, and stood higher at its shoulder than most basketball players. Its domestication occurred 8,500 to 6,000 years ago, coevolving with humans either within North Africa or introduced from Asia. Many believed humans tamed the aurochs through salt licks and thirst, making it increasingly submissive in the face of reliable water.

Domestication carried risks, but what a reward! Cattle embodied a mobile life insurance policy, meat locker, status symbol, bank account, tractor, military assault vehicle, and leather goods warehouse on four legs.[15] Manure enriched barren landscapes, but cattle also brought human fertility. Through bountiful milk supplies, lactose-tolerant herders enjoy astounding advantages over hunter-gatherers. With no animal milk, hunter-gatherer mothers like Qoroxloo breast-feed for longer periods of time—sometimes several years—before children can process solid food. Until then, nursing acts as a contraceptive. By contrast, cattle clans can wean earlier, give birth more often, and raise more children to maturity, generation after generation. As Gandhi argued more accurately than he knew, "the cow to me means the entire sub-human world, extending man's sympathies beyond his own species . . . she is the second mother to millions of mankind."[16] Even without force, sheer reproductive success alone enabled herders to overwhelm hunter-gatherers. Over several thousand years, as the Sahara dried up, early Bantu herders could sweep out from West Africa and overrun the continent, stopped only by pockets of disease and aridity.[17]

Parasitic disease erected a wall against cattle. Africa's twenty-two species of endemic biting, scissor-winged, blood-sucking tsetse fly guarded the wild humid landscapes by infecting livestock with deadly tripanosomiases, or Ngana.[18] In moist savanna and woodlands where tsetse thrived, cattle could not. Also, because cattle must drink every day, dryness raised an equally formidable barrier. Throughout most of southern Africa, freshwater remained too erratic, too scarce, too well concealed, or did not gather long enough to allow sedentary cattle. Herders lived nomadically, or on the margins, but for millennia cattle could not be herded into the lion-dense sands of the dry Kalahari.

Three technologies irreversibly broke down these last barriers to entry. First, aerial spraying of DDT killed off tsetse flies in humid lands. Then barbed-wire fences redefined property ownership and encouraged cattle proliferation into Bushmen territory. But DDT and barbed wire couldn't produce water in drylands; that required the heavy machinery of the industrial age.[19] By drilling down deep beneath the earth's surface to tap and extract centuries-old groundwater, the borehole broke dependence on rain.

Boreholes punctured the earth like pins in a pincushion, spreading fastest to the driest regions until they irrigated more food than rivers. People rightly praise Norman Borlaug's high-yield grain varieties, synthetic fertilizers, pesticides, or herbicides for increasing food productivity, but the global Green Revolution owes its deepest debt to boreholes. Boreholes banished famine, fed India's billions, and now Africa demanded its turn. So in the Kalahari, where 83 percent of all rain evaporated, 14 percent transpired through plants, and 2 percent ran off into ephemeral trickles and pans, Botswana drilled and tapped that last 1 percent in an almost biblical act. The government struck barren sands with a mighty steel rod to release a miraculous fountain. In 1903 a geologist drilled the first borehole in Namibia; even by 1950 only a few dozen boreholes had been drilled in Botswana. But over the next five decades the country punched deep straws into the Kalahari to suck up 24 trillion gallons. Some twenty-one thousand boreholes were for the exclusive use of cattle posts surrounding the reserve; only one quenched the thirst of Bushmen inside.[20] Then, even that one was permanently sealed off.

As the cattle grew fat and abundant, few saw any downside to boreholes. Then, gradually, a threefold reckoning ate into the foundations supporting Botswana, Mexico, India, China, and even the United States. One issue was costly maintenance. Tipping the scales at one ton per 265 gallons, water weighed more than oil. Lifting water from deep beneath Botswana burned up equipment and exceptional quantities of fossil fuel, leading to breakdowns and expenses that brought pumping to a halt. Today, thousands of Botswana's rural boreholes sit broken, rusted, clogged, and corroded from neglect. Average boreholes last thirty years; Botswana's died within ten.[21]

Second, even success brought risks. Given enough cheap fuel, tapping invisible groundwater was like grocery shopping on someone else's credit card; eventually the account dries up. Botswana pumped fossil aquifers for Kalahari cattle, mining ancient water 900 percent faster than rain could replenish.[22] Soon nothing was left. The addiction was universal: America's subsidized midwestern farmers, knowing the limit is near, race each other to the bottom of the Ogallala Aquifer until their wells run dry.

Finally, politics entered. In a dry land, water means wealth and power. In theory, boreholes should decentralize authority over water use and control, and bring democratic decision making to smaller units more autonomous than centralized federal dams. In practice, the elites control groundwater as well.[23] Government owns half of Botswana's boreholes; select cattle barons corral the rest. Even when charities drill water for marginal tribes like Bushmen, their boreholes have been quickly appropriated by well-connected ranchers. Cattle posts consolidated a feudal culture in which owners granted access to water in exchange for physical labor or sexual favors. As Botswana sank boreholes to expand cattle ranching, foreign academics watched this dynamic play out before they could complete research on their dissertation. The central Kalahari held out longest, but for most of her life Qoroxloo occupied a dry island in an encroaching continental sea of cattle.[24]

The thirst of those surrounding cattle was so powerful that it might even affect groundwater beneath the reserve. Like a straw in a milkshake, all those borehole pumps sucked down water in a circular "cone of depression" that lowered water tables, perhaps permanently, for miles in all directions. Botswana ranchers claimed groundwater was "unlimited" and "flowed" like an underground glacier that no amount of pumping could use up. Scientists countered that rains never seeped down more than twenty feet deep; everything below that was finite "fossil water," and cattle post boreholes were depleting the Kalahari's shallow wells, springs, and aquifers.[25]

Such a claim sounds preposterous, until one considers not only how much beef people consume, but also how much water cattle consume. It turns out a pound of beef requires 5,200 gallons; in its lifetime the average dryland cow requires 1.5 million gallons.[26] Provided with subsidized groundwater, Botswana's herd expanded exponentially from a few hundred thousand cattle in 1950 until today 2.5 million cattle populate the country surrounding the Kalahari reserve, annually consuming 3.75 trillion gallons of water directly or embedded in feed.[27] Negative impacts were inevitable for any arid country dependent on groundwater. As midwestern ranchers have pumped 13 trillion gallons of water from the Ogallala Aquifer, for example, they have dried up wells in Kansas, Nebraska, Oklahoma, Colorado, and New Mexico.[28] The irony of the Kalahari was that fewer and fewer Botswana citizens outside the reserve owned cows, herded cattle, or could even afford meat.[29] Instead, under lucrative subsidies like water, Botswana's Meat Commission exported each slab of beef out of the country, along with the small reservoir of water required to produce it.

Cattle inhaled water supplies in one other insidious way. In 2002, some 1.5 billion cattle ruminated worldwide. As they did, microbes respired in their gut through an anaerobic process that produced methane gas and nitrous oxide, two invisible contributors to global warming respectively 23 and 296 times more potent as greenhouse gases than carbon dioxide. Cattle emit 95 percent of their methane not through farting but by burping: Each day, cattle collectively generate 18 percent of Earth's greenhouse-gas emissions, worse than all the cars driven all over the planet, combined.[30] One writer quipped, accurately, that, "given the amount of energy consumed raising, shipping and selling livestock, a 16-oz. t-bone is like a Hummer on a plate."[31]

In the Kalahari every action produces an equal and opposite reaction. Each fraction-of-a-degree rise in temperature brings a protracted dry spell that reduces Botswana's herd in a climatological culling process. In the early 1980s, drought killed off a third of Botswana's cattle. Aggressive borehole drilling built back the population, only to see another massive die-off in the early 1990s. These droughts were invariably described as "the worst in living memory" or "the most devastating in the last century." Yet 2002 brought another, from which the country's stagnating cattle industry has not recovered. By obediently following Western examples, instructions, and export quotas, Botswana's government subsidized an outmoded way of producing meat that failed during drought, destroyed groundwater reserves, spread disease, sickened consumers, killed wildlife, and accelerated the desiccation crippling the land.

OUR HUMAN QUEST for meat will continue to escalate.[32] For every newly converted vegetarian, four poor humans start earning enough money to put beef on the table. In the past three decades, the earth's dominant carnivores have tripled our average per capita consumption; in the next four decades global meat production will double to 465 million tons.[33] The question is whether we will convert water into wild or domesticated protein. If past is prologue, governments will continue to eradicate diverse and water-adapted game species, install thirsty European livestock, bring intransigent local people to their knees, and extinguish the age of the hunter.

Three short generations ago, half of North America roamed the backwoods armed with rifles seeking prey and providing meat for the pot, but their ranks have plummeted each decade,[34] sinking to a historic low today of barely 12.5 million,[35] or roughly 4 percent.[36] The only force that may be powerful enough to reverse that downward spiral is aridity. Nations that demand affordable meat increasingly will be forced to shift their hunger back toward drought-tolerant and climate-friendly game species.[37]

For better or worse, the old instincts never entirely die out. They linger long after people are torn from a home, or restrained from venturing out with spear, bow, and arrow. Though driven from his Kalahari homeland, Mogetse Kaboikanyo was reluctant to accept dry foodstuffs doled out by the cattle-loving government in exchange for Bushmen submission. "Our future comes from the lives of our children," he said. "Our future is rooted in the hunt and in the fruits which grow in this place. When we hunt, we are dancing. And when the rain comes it fills us with joy. This is our place, and here everything gives us life."[38] He echoed Sitting Bull, who said, "When the buffalo are gone we will hunt mice, for we are hunters and we want our freedom."[39]

Qoroxloo's grandsons wanted their freedom. They must have wanted it very badly, given the danger of heat exhaustion or predator attack; given the threats of arrest, fines, jail, or years in prison; given the constant menace of being beaten to death; given opposition by vegetarians, animal rights activists, wildlife ecologists, safari hunters, and wilderness purists. Moeti could not easily convey the high-risk drive in his heart, but hinted at it through his stories and silences. In the dry world, precious few possessions endure long enough to be passed down across generations. No cattle, no barn, no house, no stock portfolio, no trust fund, no watches or wardrobes or sofas. Only the process, the intimate ties to the surroundings can be bequeathed as a birthright.

The brothers spoke of that birthright as they sat around the fire, listening to the stars. The dogs lay flat, sated from gorging on antelope intestines, and the swarms of flies vanished as temperatures plunged into cool night. As the flames subsided, Moeti recalled a boyhood of long treks spent alongside his grandmother, Qoroxloo, trying to earn her confidence and build his self-knowledge as she taught him to identify the tracks of small and large prey species. In one of his earliest memories, he sat on his father's shoulders and saw a group of relatives return through the bushes and trees. Among them was Qoroxloo, returning to Metsiamenong from a successful big-game antelope hunt. Not all women hunted with the men, but Qoroxloo did. She liked to accompany young men on hunts, and she was good at it, and helpful. She did not touch the bows, arrows, and spears. But she helped track the animals, and she knew when to keep silent and only use hand signals. Qoroxloo's presence on hunts may seem unusual, he said, but she "depicted a complete naturality and a cultural transfer from generation to generation in line with the spirit of Kua nqaai."

The fire died down to glowing embers, ceasing to spark. Before drifting off to sleep, Moeti finished that story of his earliest memory. Qoroxloo strode back home into the camp, where as usual she helped butcher the meat to distribute and share and cook in preparation for the dance. He could not know whether his grandmother's apparent ecstasy came from reflecting on a successful hunt or anticipation of a celebratory dance, only how she had the blood from the animal on her hands, and blood smeared over her shoulders from where she had draped it. She was smiling through all that blood, Moeti recalled. She was smiling brightly, and she was singing.

CHAPTER 10

Survival of the Driest

BESIEGED PEOPLE WALK AN INVISIBLE LINE between courage and folly; at some point stoicism blurs with suicide. Witness the defiant last Mohegans, Civil War rebels, Australian Gallipoli grunts, Kamikaze pilots attending their own funerals. For centuries, Bushmen fit into this fatal role. Surrounded, outgunned, outnumbered, they never surrendered.

Qoroxloo's last stand appeared to be a gallant yet tragic finale to previous centuries of Bushmen decline. Her rejection of government water seemed as rash as a Christian Scientist's rejection of penicillin. Sure, they could reduce losses by minimizing energy expenditure, and squeeze enough water from plants and animals to secure metabolic functions, but dryland survival rose beyond nutrition and quenching thirst. Every child who has been potty-trained or marched to the bath knows freshwater brings cleanliness, godliness, and human progress, which helps explain why governments sing the universal mantra in a global chorus: "Water is life!"

Evidently Qoroxloo understood water as a bit more complicated. She rejoiced in the wild rains that replenished animals and plants. But as the government reconfigured, mobilized, ordered, and domesticated this water, it became something different. Once tamed, water appeared to introduce disease pathogens and transported invisible microbes, chemicals, and processes that could prove dangerous, even fatal. It wasn't water itself that Qoroxloo rejected, but those unwelcome consequences and over-dependence that accompany government provision of it.

Nor was she alone in her suspicions. Six months into the siege, as the drought worsened, hundreds of Bushmen tried to sneak past the reserve boundaries. Young men crossed the threshold in search of meat; grandmothers on the verge of collapse headed off to reach security; one woman with two children traversed the sands on foot, at one point tracked by lions, to provide her family with a better life. All these Bushmen came to realize that they had endured enough hardship, and finally surrendered to cross that "line in the sand," just as President Mogae had predicted. What he hadn't expected was what direction they'd be going. Rather than departing for "civilization," his so-called Stone Age creatures had begun leaving the bleak government resettlement camps and heading back to their homes in the Kalahari, voting with their feet.

The Bushmen exodus defied both logic and expectations. Development experts sent in researchers, foreigners dispatched TV crews, and President Mogae set up a special task force to grasp this phenomenon and determine why Bushmen were going back in and returning home.

I brought an Australian Broadcasting Corporation crew into the Kalahari to investigate what was going on. For several days in the reserve, they saw water from the Bushmen perspective, inside out, testing assumptions about dry sanitation, dry hygiene, dry health, and survival.

First they considered dry sanitation. The end of government water in no way altered how Bushmen answered nature's call. Like the rest of her remaining band, Qoroxloo discreetly and regularly headed off from camp a few times each day. The reserve lacked indoor plumbing,[1] but Qoroxloo isolated urine to use the resulting ammonia for curing the hides of animals she snared, found privacy and seclusion in the endless shrubs, and chose leaves that provided soft, organic, biodegradable toilet paper.[2] What's more, from an early age Qoroxloo had been handling messy feces—her own, her children's and grandchildren's, and those of the elderly after sphincter and bladder muscles weakened—releasing or carrying excrement out into the Kalahari. As the film crew discovered, there was nothing terribly romantic in this. The moisture released from within bowels and bladders sent out a signal like a beacon across the Kalahari so that, almost before waste hit the sand, flies swarmed in out of nowhere to alight. Yet Qoroxloo's band never seemed to regard their raw excrement as the deadly, debilitating, chronic threat that so plagued and decimated all densely populated cities throughout time. By constantly dispersing, radiating outward, away from the epicenter of settlements, the Bushmen's decentralized existence combined with dry heat to bake off disease vectors before they could spread. The hot dry sand—like a giant swath of lion litter—absorbed and desiccated the moist waste, quickly killing off harmful organisms.

Next the Aussies weighed the pros and cons of dry hygiene. Like most Bushmen, Qoroxloo struggled her whole life against filth and flies, and enjoyed washing, when given the rare thundershower. But in dry seasons, and when she was denied water during the siege, Qoroxloo cleaned cookware, knives, utensils, and pots by scouring out the interior with sand, and she would also scrub babies and children as vigorously as possible. Her grandson, Galomphete, described how after eating a tsama melon, Qoroxloo would grind and break the wet melon seeds into smaller particles, add in the leaves of certain plants that produced a soapy mucilage, "mix them with a little bit of water, then take these seeds and wash me by rubbing them against my body until I would become as clean as a person who uses water for bathing."

The reserve was in no way quarantined. If anything, Kalahari Bushmen were, like pre-Columbian Native Americans, particularly vulnerable to certain viral infections to which outsiders had developed immunity: Flu could be fatal; some still bore the scars of smallpox that swept through the Kalahari in the 1950s; and venereal diseases moved from band to band. Yet by forcing them to disperse often in small autonomous bands, aridity in the hot empty spaces provided a certain ecological "drywall" that lowered their risk of exposure to and transmission of deadly infectious diseases endemic to tropical Africa, diseases largely related to water.

As missionaries noted, no "fever trees" grew in the central Kalahari; in only the rainiest years did malaria or "bad air" sweep across the desert savanna, with minimal impact. Cholera was unheard of, as was arsenic contamination. No standing water could support giardia, and no stagnant ponds lingered long enough to allow bilharzia snails to reproduce and work their charms through the skin of bathing humans. Mosquito-borne encephalitis and Rift Valley Fever also were unknown in the arid landscape. In this respect, the end of water delivery unwittingly reinforced the Bushmen's arid insulation. Qoroxloo could capture the benefits of water found in her food or sealed beneath the sands, yet avoid or escape the worst aspects of standing or flowing water: insect-borne pathogens, bacteria, viruses, chemicals, and pollution. Small wonder humans evolved in arid tropics, since "the desert is also a healthy place to live. There was no tsetse fly, no malarial swamp, no raging cold and chilling wind. Because everyone was out of doors much of the time, the spread of disease was much less of a risk than in colder climates."[3]

This brings us outsiders, third, to confront issues of dry health and survival, measured either by infant mortality or by longevity. After Botswana cut off water essential to life, it also halted mobile clinics designed to stave off death. The risks of childbirth or death from severe wounds were real in the Kalahari, but if Bushmen wanted obstetric/gynecological support, first aid, anesthetic treatments, or curative medicine, they had to seek certified doctors, outside the reserve, or find amateur Qoroxloo, at its heart.

In the middle of the dry season, during the siege, the newly pregnant Tschaekom was anxious about the mysteries of motherhood, but her great-aunt alleviated day-to-day concerns. As an elder and midwife who had assisted with dozens of births, Qoroxloo knew the benefits of nearly a hundred wild dryland medicinal plants, which concentrated oils, some of them toxic, to repel herbivores. Over time, as Bushmen learned from elders what plants could help or hurt them in which ways, the Kalahari became a pharmacopeia for those experienced enough to read the unwritten labels and dosage instructions. Qoroxloo knew which plants aided conception, relieved menstrual cramps, or deliberately brought on contractions to abort an undesired fetus.[4]

Tschaekom desired a baby to raise in the Kalahari, so in preparation for that event Qoroxloo began gathering plants: Saponins in wild cucumber would help relieve the pain of childbirth; fiber from "mother-in-law's tongue" would constrict hemorrhoidal tissues; leaves of devil's thorn, when mixed with moisture, would produce a slimy mucilage that served as a childbirth lubricant; and the vicious devil's claw could be applied as an ointment against a difficult birth or boiled as an infusion to treat postchildbirth diarrhea or infertility.[5] The desert aridity ruled out luxurious Western "water births," and Bushmen delivered babies off in the bush, squatting over hot sand.[6] But an experienced midwife like Qoroxloo ensured Bushmen birth rates were average and infant mortality was moderate, until eventually rates of survival to adulthood even rose slightly higher than those for Botswana citizens outside the reserve.[7]

That said, dry health remained precarious, and no one knew this better than Qoroxloo. Days before the Australian TV crew arrived, her husband, Mohame Belesa, had stepped on a broken branch, which sliced through his thick sole and penetrated to muscle tissue. He began bleeding profusely and pressed his hand over the skin to hold in the precious fluid. It eventually clotted, but larger risks remained. In the Kalahari, hyenas, lions, and even leopards mostly leave healthy Bushmen alone but may single out the old, sick, and vulnerable. More deadly are all those tiny invisible microbes that come in the night to fester and rot the flesh. As an antibacterial and antibiotic, Qoroxloo knew to dig up and apply roots from the shade-growing tattoo plant,[8] but as days passed his unhealed foot threatened to turn septic.

Abundant water might have helped bathe, soothe, and irrigate her husband's suppurating wound, and sanitary bandages with Western antibiotics like Neosporin ointment could—and when borrowed from the TV sound engineer, eventually did—stave off infection. But Qoroxloo continued to watch her husband hobble, read the pain he suppressed, and grew anxious. She checked his wound as often as he let her, knowing that if gangrene set in, she might have to use a knife to sever his foot without anesthetic. That prospect appeared horrific,[9] but preferable to his surrendering and departing, with or without her, to a clinic in the odious New Xade, from which he might never return or fully recover.

Qoroxloo and her band seemed to intuit what researchers have only begun to prove: Healing can be as much psychological as biological. In fact, experts found that "provision of health care in squalid resettlement camps" never compensated indigenous people for "the denial of a lifestyle that is central to their concept of health and well being."[10] Whether traditional cures or Western pharmaceuticals, medicine requires confidence in a larger context or association. Central Kalahari Bushmen take strength from place. They require proximity to the graves, and lands once inhabited by the deceased ancestors. "If I'm sick," said Dauquoo Xururi, "they're my hospital."[11] Many described the power of place, of the land's spirit, to heal. "We need sand from near their graves to mix with plants and make medicine. That is our way. We still live with our ancestors. Our children have to bury us, and their children have to bury them, in our land. That is how it is with us."[12]

Such beliefs are by definition impossible to test in a laboratory or clinic, and yet given equivalent diseases, patients do tend to recover faster and more completely in familiar places, surrounded by loved ones, smelling familiar odors and hearing sounds they grew up with. Conversely, psychological stress brings physical sickness. Once Bushmen were evicted, their health rapidly deteriorated, largely, said researchers, because "Bushmen define themselves in terms of their relationship with their land, making their environment essential not only for physical provisioning and regulating services, but also for physical and cultural survival."[13]

On our way out of the Kalahari, with that Sydney-based TV news crew, we passed, and filmed, dozens more Bushmen headed back home, laughing and singing. The Aussies continued to debate the dissident Bushmen's logic and admire Qoroxloo's defiant will, but upon departing the reserve these urban whites raced to the nearest hotel to rejoice again in flush toilets, languish in deep soaking baths, scrub sand from hair and skin, and rest assured in the knowledge that a hospital was nearby. No one wished to trade places, to cross back to a world of dry sanitation, dry hygiene, dry health and survival. But in an increasingly hot and dry climate fewer humans would enjoy that luxury of choice.

The question, then, was which approach worked best in an age of permanent drought.

COMPARE QOROXLOO'S WORLD with the outside world's preferred alternatives, starting with our gloriously wet sanitation. Kalahari Bushmen might find it curious not only how we Americans collectively blend 300 million quarts of urine and 100 million pounds of feces into perfectly good drinking water, but take such immense pride in our daily feat that we expect the rest of the world to follow our example.

UN development standards and $8 billion in donor incentives all but ensured that leaders like President Mogae would race to adopt and import our porcelain technology as a matter of dignity and pride. So now, outside the Kalahari reserve, half of Botswana sits smugly enthroned. Expensive flush toilets sweep away each daily emission with six gallons of freshwater gushing through pipes to septic tanks and treatment plants, stagnating until diluted with yet more water. This septic stew has generated unnecessary evils. First, wet sanitation actually helped prolong the lifespan and expand the distribution of intestinal parasites by providing them with a warm, soupy habitat to breed and grow. Worse, as the arid country dried up further, Botswana officials continued subsidizing the installation of hundreds of thousands of prohibitively expensive toilets and pipes and sewerage to people who increasingly lacked either the cash or the water to flush.

As cities like Gaborone exploded from 20,000 to 200,000 over the course of a few decades, the sewer system became overwhelmed by numbers and sprawl. Rural immigrants answered nature's call as they always had, but with concentrated density, the urban trickles of runoff and puddles became open sewers, with the predictable spread of waterborne worms, cholera, dysentery, and diarrhea. This problem went beyond southern Africa; worldwide, nine out of ten people discharged raw sewage into rivers and lakes without treatment. Contamination of clean water killed 1.8 million people—two per minute, around the clock[14]—a number eclipsing deaths from AIDS, malaria, and all wars combined.[15]

Under climate and demographic pressure the situation was expected to worsen. Optimistic scenarios, such as the unlikely event that the UN Millennium Development Goals would be achieved,[16] left 34 to 76 million people (mostly children) dead over the next two decades.[17] Experts ranked bad water "the most serious public health crisis facing humankind," and "failure to provide safe drinking water and adequate sanitation as the greatest development failure of the 20th century."[18]

President Mogae evinced little interest in other systems, especially those pioneered by Bushmen. But their Kalahari sanitation principles had given rise to a decentralized, convenient, off-the-grid technology that could liberate billions who wanted indoor plumbing but lacked disposable water or disposable income: the urine-diversion compost toilet. It resembled any toilet but, as the name suggests, diverted urine into a container while feces fell into the pit below, where soil or ash or sand helped dehydrate the feces and kill off pathogens. No smell, no flies. Dried-out human waste could be used as manure. South Africa's austere provinces embraced this revolutionary urine diversion approach, as did Asian cities ringed by slums and even affluent homes in Europe. With no flush, nothing to dispose, no need to monitor or replace aging pipes and leaky tanks, Kalahari Bushmen own and operate the oldest, cleanest, cheapest, safest, and simplest dry sanitation technology available, avoiding the high risks of waterborne and sewage-related disease while saving every precious drop.

In the rich world, few of us would willingly forgo hot showers in the name of better hygiene. Yet our warming world of mass poverty, permanent drought, and acute water scarcity reminds us that, even for those who can afford abundant water, no perfectly safe options remain. Our current "clean" culture may have emerged as a relatively recent, and extreme, phenomenon; for millennia even kings used to bathe no more than once or twice a year.[19] Now researchers say obsessive-compulsive Westerners may in fact be getting "too clean." Natural dust, sand, grime, even toxins and germs are rarely the absolute evils portrayed in soap commercials. Sand is not dirty. To the contrary, long-term immunity grows aberrant or compromised for children raised in a dirt-free, germ-free, antiseptic household where vacuum cleaners and Windex expunge exactly what is needed to build up resistance.[20]

More devastatingly, in the developing world the expansion of new water infrastructure into arid lands brought with it a dark side that few officials have cared to acknowledge. From malaria and ebola to ringworm, typhoid, yellow fever, dengue fever, schistosomiasis, and sleeping sickness, Africa never lacked invisible killers, only vectors through which they could spread. These deadly diseases all migrated with the insects—tsetse fly, ticks, snails, worms, and mosquitoes—that bred in and congregated around water. Like other dry expanses, the Kalahari offered scattered bands of human settlements an evolutionary advantage that infrastructure development could reverse. Government dams, wells, and open tanks provided a Petri dish irresistible to the natural profusion of microbes and insect larvae, pestilential organisms that now had sufficient concentrations of water and people to incubate, gestate, and reproduce.[21]

This law of unintended consequences meant that even well-meaning efforts to "improve" water supply to "backward" dryland herders, semi-nomadic pastoralists, and dispersed hunter-gatherers all too often worsened their lives. Global crises in malaria, sleeping sickness, bilharzia, and other waterborne diseases could be perversely magnified by the cure: health-oriented water projects.[22] As speed and scales rose, so did impacts; one irrigation project required $20 million a year just to counter the waterborne disease it had introduced and spread.[23] To minimize devastation, health officials urged safer water storage designs, or halting new projects until more was known. The precautionary principle of global water development mirrored the Hippocratic Oath: First, do no harm.[24]

Instead, Botswana plunged ahead. It drilled and pumped groundwater from beneath the Bushmen's ancestral homes, then piped water outside the reserve to the new relocation center at New Xade, whereupon reliable government water supplies generated a portfolio of maladies. The predominantly male development workers with disposable income soon spread HIV/AIDS among newly idled, cash-hungry Bushmen females. Next, stagnant pools and mud puddles around running taps attracted clouds of blood-sucking, disease-spreading parasites. Then a lethal epidemic broke out in which hundreds of relocated Bushmen sickened within weeks of each other, and fifteen died quickly after vomiting, and breathing difficulties.[25] Panic spread, as evicted Bushmen traced their ailments to the foul-tasting water supply. Botswanan authorities never sent a doctor to investigate the cause and shrugged off the consequences, but these cases reinforced how the sudden careless infusion of water into dry, dense landscapes could prove riskier than water's sudden removal.

No wonder Bushmen dissidents found the health situation outside the reserve so dangerous. Tshaekom preferred to give birth in the Kalahari rather than a sterile, water-abundant clinic, and when foreign reporters asked "what made her take the risk," she stared back in disbelief; ever since it was created in 1997, New Xade was known among all Bushmen as "a place of death." When they first saw it, the Australian TV crew, like many foreigners, thought the resettlement camp looked rather orderly and neat, albeit soulless. Only upon a closer look could people notice the impoverishment, marginalization, subjugation, dependence;[26] the diabetes, iron deficiency, spousal beating, and lack of protein.[27] Even the government conceded that New Xade Bushmen were dying from alcohol poisoning and cirrhosis of the liver,[28] and a troubling new sexually transmitted disease that spread quickly and killed slowly. STDs were nothing new to Bushmen,[29] but unlike two out of five Botswanan citizens, those inside the reserve had avoided AIDS until evicted and concentrated in government camps. "We only expect old people to die," lamented Roy Sesana, "but now the young are dying like never before."[30]

Over the years Qoroxloo brought dozens of healthy infants screaming into the central Kalahari, several of them born after the government cut off water. Tschaekom decided "it was no worry for me, without water. This was the kind of life I was born into." Two decades earlier she herself had been brought into the world under Qoroxloo's ministrations. Now in the middle of a dry winter, Tschaekom gave birth to a baby daughter, Kebadumetse, who exercised healthy lungs and who a year later still appeared to be in ruddy health. In the outside world water-abundant New Xade became known as a place of death; inside, Qoroxloo could ensure the arid reserve remained a place of life.

The two worlds would collide as both went after the same supply of water.

PART III

Competition

CHAPTER 11

Water for Elephants Only

WHEN BOTSWANA'S MINISTER OF Local Government, Margaret Nasha, cut off government water to Qoroxloo's people, her intent was not personal. The last bands of Gwi and Gana in the Central Kalahari Reserve had not been singled out for unique discrimination. The rest of Botswana's fifty thousand Bushmen[1] had already been relocated, evicted, or dispossessed in the last few decades, many of them from other parks, forests, and reserves. Apparently it had to be this way because, as everyone knew, these places were exclusively for wild animals, not people.

Qoroxloo wondered why people could not live with the animals as they always had.

But Nasha repeated that the reserve needed to be depopulated for the greater common good, and by eradicating people from wild places Botswana was merely conserving nature just like in those protected wilderness parks first established by the United States of America.

Qoroxloo's cousin, Nyare Bapalo, didn't believe Nasha and told her so to her face. People are part of the game, he explained to me. "At times we die and the animals eat our bones, and other times the animals die and we eat their bones." Bapalo had been away hunting during the final water cutoffs and returned to find that his wife and son, caught in the uncertainty, had been trucked away. But if he left now he knew he could never return with them. "We will not go," Bapalo said. "If the government returns, we will sit here. If they kill us, then they will kill. This is our home. We are the first people in Africa. We will stay. Though they take away the water, we will stay."

They stayed. But despite taking away Bushmen supplies, Botswana guarded another borehole elsewhere in the Kalahari for the sole use of wildlife, and in so doing officials exacerbated the region's inter-species conflict over water.

One day during a hunting excursion three men from Qoroxloo's band came across tracks that generated anxiety and excitement. To Bushmen, no two animals leave exactly the same prints, and each reveals a distinct individual. This one left behind: two pairs of matching ovals; larger, deeper pairs spaced apart at some distance; a bit of sand kicked up on one edge. Each oval showed a pattern of nooks and crannies as unique to this animal as fingerprints are to humans. If dead, this unfathomably heavy animal could feed their band for months,[2] but if still alive, the solitary beast could instead kill them. This lethal, near mythical creature, described in Bushmen stories but rarely seen this far south in the Kalahari, revealed either their salvation or their doom. To predict which, and to understand behavior, Bushmen learned to project themselves as intimately as possible into the outlook of animals whose spoor they followed, and the life and fate of this one most closely mirrored their own.[3]

DARWIN CHOSE *LOXODONTA AFRICANA* to illustrate his evolutionary theory, for the elephant exemplified natural selection. Despite its long gestation and low reproduction rate, in five centuries a single breeding pair "would produce a standing population of 15 million individuals."[4] Yet, even as *On the Origin of Species* was being printed, Africa's elephant numbers were already significantly lower. A century later they began shrinking even faster.

For decades the West blamed the decline on poachers. Greedy ivory traders, wantonly gunning Jumbo for tusks, made great villains. Their lurking presence brought donors to fund government forces, who waged armed warfare through helicopter gunships flying on shoot-to-kill orders, and this international mind-set peaked in 1989 as Dr. Richard Leakey symbolically set fire to a twelve-ton pile of confiscated ivory tusks in what he called the climax of his bold and controversial work to save the elephant in Kenya.[5] Unfortunately, "poaching" was never conclusively shown to cause serious decline. Even as ivory's value rose from $100 to $350 per pound in one year and authorities seized a record twenty-four tons destined for Asia, those twenty-three hundred tusks represented 0.4 percent of the estimated wild African elephant population. That was far less than elephant mortality; meanwhile, populations annually grew 7 percent. Scientists traced confiscated ivory through DNA to a place of origin but could not say whether tusks came from killing or were collected after elephants died of natural causes.[6] Disease killed a few hundred elephants each year. Lions and hyena dispatched dozens more. Yet the underlying cause whenever elephant populations plummeted was traced largely to thirst during protracted droughts.

Elephants must drink eighty gallons of water at least once every other day. Thirst drives daily and seasonal movements in a ceaseless shuttle outward from dwindling water sources toward ever more distant browse. Thirst kills old, sick, and young elephants. Four out of five elephants that die during dry spells are younger than twelve. Drought expands the distances between hunger and thirst. Vegetation withers, and pans dry up. After a year-long protracted dry spell in Kenya, for example, thirst slashed elephant populations by a quarter.[7]

Somehow, elephants sensed the end of water in advance. The effects of thirst inhibited reproduction; during drought fertile females proved unwilling mates and gave birth to fewer calves. That compounded tensions among already irritable males. Elephants drove other creatures from vanishing water holes, as they dug deeper in dry riverbeds until the water table dropped below the reach of their trunks, whereupon they panicked, forced to move long distances under duress. Drought had most likely driven this elephant across the Kalahari, wracking his impressive memory to recall the last place he might have quickly found eighty gallons of water: in some pan or fossil river valley. But with the water cut off, his options were limited to one artificial wildlife puddle or an ecotourist lodge swimming pool outside the Kalahari reserve. Both lay a hundred miles away, farther than the average elephant could walk in two days before it died of thirst.

Qoroxloo's camp lay closer.

So where foreigners saw a gentle giant, the Bushmen saw a parched and relentless terminator. If he smelled food and water back at Qoroxloo's camp, the elephant would not leave Bushmen alone, no matter how much noise they made, or how many flaming wood pieces they hurled. He would sniff out hidden water and splinter and shatter anything holding it. He would eat their last tsama melons and guzzle their last water, not in cruelty, merely to survive at their expense.

He wasn't the first to seek water here, nor was he acting alone. If he failed, more would come. Over the last two decades one thousand elephants congregated along the Kalahari's western edge, where few if any had lived before, overcrowding the Bushmen's water supplies. Each day those elephants needed forty thousand gallons.[8] To get it, one elephant had covered a borehole and inhaled water up from a depth of sixty-five feet. At night you could hear the elephants fighting for hours over a single water point. They sniffed out boreholes, knocked over mills, tore up pumps, and uprooted wiring. They pushed over trees, climbed up and over stumps, stampeded through thick wired and electrified fences. Elephants literally brushed human rivals aside with a fatal nose press or a side kick. Poisoned arrows bounced off skin two inches thick; a spear stab risked enraging it into a brain-crushing whirlwind. No thirst-driven wild animal, not even rival humans, more threatened Bushmen survival.

And none had more powerful allies.

OUTRANKING THE PANDA bear, the sea turtle, and the timber wolf as an emblem of nature conservation and Wilderness, the African elephant inspired global Greens to no end. His photos and profiles permeate the glossy documents and Web sites of big nongovernmental organizations (BINGOs) like Conservation International, the Nature Conservancy, the International Union for Conservation of Nature, the Wildlife Conservation Society, and World Wildlife Fund, among hundreds of pressure groups. He is an animal-rights darling and antitrade emblem. International entities deploy him to move legislation and invest billion-dollar budgets to buy up and enclose landscapes the size of small countries for his habitat, a top-down tactic known as *fortress conservation*.[9]

In fairness, nature emphatically needs defending. Our crowded world is frenetically stripping, shrinking, exhausting, despoiling, exploiting wildlife habitat, and wiping out charismatic species like elephants at unprecedented rates. Who could oppose "saving the last great places"? Who denies "extinction is forever"? By thinking locally, acting globally, the strongest organizations in Europe and the United States zoomed in on "virgin" forests, "pristine" wilderness, "unspoiled" coral reefs, and "ecological hot spots." Over the course of a century they fenced off one ninth of the earth in "protected areas," adding up to an area larger than Africa.[10]

For much of that conservation century, water remained something of an afterthought. Before rivers flowed into these ecological fortresses, dams cut up the waters like a liquid pie, with big slices for farms, energy, cities, transportation, industry, and recreation. Pumps divvied up aquifers the same way. But soon aquatic biodiversity declined much faster than all other ecosystems, as two in five species have become threatened with extinction.[11] As fisheries crashed and parks and refuges suffered, Greens counterattacked, demanding, as they had with land, that a certain amount of water be set aside "for nature." An entire scientific discipline grew up around the near-existential question of determining how much water a stream required for itself, with ecohydrologists' requisite "in-stream" or "environmental" flows determined on the needs of all wild creatures great and small. As nature's charismatic ambassador, the elephant fused terrestrial and aquatic habitat. It ranged across national and international borders, demanding more and more of wild land and wild water in Africa's ecological hot spots. It has become the ultimate *umbrella* species: Secure the elephant, and you lock in thousands of smaller species as a bonus.

Even better than his ecological value, the elephant was economically sustainable, generating revenues for fortress lands and environmental flows. As market-based conservation dictated, "If it pays, it stays," elephants became the ultimate cash cows.

Every overseas visitor to Africa understandably had to see, hear, and photograph one. With sightings almost guaranteed, Botswana's tourist trade roared. The country's $486 million in direct ecotourism revenues grew 5 to 7 percent annually—as fast as the pachyderm's population—while elephant-based conservation fueled $1.6 billion in broad economic activity, generated 10 percent of all jobs and 16 percent of nonmineral GDP.[12] Botswana took pride in its reputation for high-yield, low-impact conservation. By helping net $500 to $1,500 per visitor, per day, Earth's largest land mammal was a top money maker. Not only did it pay and stay; it doubled each decade just as Darwin had predicted. So while elsewhere in Africa farms confined elephants to islands surrounded by a sea of humanity, Botswana invested in the opposite. It hosted a fifth of the continent's elephants. In the northern Kalahari sandscapes, 120,000 *Homo sapiens* were surrounded by 120,000 *Loxodonta africana* in a one-to-one ratio that made water conflicts all too common, often fatal, and worse each year as the larger "charismatic megafauna" took out water tanks, drained reservoirs, devoured subsistence crops, and destroyed the only local wells. Foreign exchange-generating elephants thrived in the north; dissident Bushmen clung to the south. The two competed for what scarce water remained in between, and only one species could remain secure.

To avert catastrophe, governments could choose either a soft or hard policy. It could let communities earn shares of lucrative elephant tourism revenues, so that conservation incentives outweighed complications.[13] Zimbabwe's pioneering CAMPFIRE program remained largely intact even as the nation's agricultural economy imploded; Namibia's locally managed conservancies empowered local people to benefit from both trophy hunting and photographic tourism. This "soft path" approach worked. But it required devolving authority over natural resources like water to people like Qoroxloo.[14]

Botswana recognized the clout of global conservation groups. Indeed, its influential vice president, Ian Khama, sat on the board of the biggest and most powerful Green group, Conservation International, which had staked its claim on top-down protection of vast, ecologically rich wildlands, keeping animal habitat pristine and preserved from the ravages of humankind. Perhaps for this reason, Botswana chose the hard-line approach: concentrate power and authority in a few elite hands. Its officials cleared indigenous people out of elephant-rich conservation areas, first from Moremi, then Chobe, then Tuli, and lastly, they planned, from the Kalahari.[15] Then it struck agreements at the top with BINGOs that left people out of the equation, granting tourist concessions that funneled ecotourism revenues back to the state.[16]

Even so, thirst-driven elephants ignored centralized state planning, overpopulating protected areas to the point where they could annihilate the habitat on which they depended. In the Kalahari, where water dictated the elephant's home range and use, a single bull may range twelve thousand square kilometers; a half dozen could bust the entire reserve. As herds doubled, and pressure on shared water supplies intensified, Botswana could roll back and clear out human settlements to make room for thirsty elephants, or it could cull elephants to protect water resources.[17]

At the mere suggestion of a cull, the elephant's international allies raised an outcry until the idea was postponed and then dropped. As Dr. Leakey proclaimed, species like the elephant deserve water first, as "the global interest in biodiversity might sometimes trump the rights of local people."[18]

Few in any real position of authority could grasp that a global interest in biodiversity might actually be one and the same with the rights of local people. The trouble with indigenous people, it seemed, was that they generated insufficient cash flow, especially foreign exchange from ecotourists. In the jargon of market-based conservation, Bushmen were not "economically sustainable," and Qoroxloo simply didn't "add value" to the Kalahari landscape in which she was born.

International wildlife groups publicly frowned on Botswana's efforts to uproot Bushmen from their reserve. Quietly, however, wildlife researchers and environmentalists from the United States and Europe confided to me that clearing out people would be highly beneficial for wildlife and would certainly simplify the current messy situation. This line of thought led neighboring countries like Namibia to remove native people from Etosha Park, South Africa to empty its glorious Kruger, and Zimbabwe to forbid humans from residing in Hwange Game Reserve. Internationally, Australia fenced out Aborigines, Brazil excluded indigenous people from new rain forest preserves, and India evicted one hundred thousand rural people to protect one tiger habitat. Worldwide, this particular category of deracinated people was estimated to number in the tens of millions, and they had been dubbed "conservation refugees."[19]

Only by posing as an ecotourist and paying handsomely to go photograph the wildlife could I break the siege after the water was cut off. When I wandered into Metsiamenong, Qoroxloo and the others gratefully received the provisions I had smuggled, but they really thirsted to learn about the outside forces for which they should prepare. What was the government planning? Why did it want Bushmen kicked out?

As we began talking about the water cutoffs and the state of the court case, Nyare Bapalo suddenly interrupted. With his grandson Moagi translating, Bapalo asked where I came from.

I answered, "America."

Bapalo nodded and mentioned Minister Nasha's last words to him about how this place was exclusively for wild animals. "And your government," he asked, "does it also push against the First Peoples, the wild people of the bush who own no cattle? Does it push them off their ancestral land?"

I paused, took a breath, and considered how to answer. Yellowstone had been emptied of the Lakota, Crow, Nez Perce, and Blackfeet. Yosemite was swept clean "of any scattered bands" of Miwok "that might infest it." Throughout my personal and professional life I had hiked from Acadia to Olympic parks, I had rafted Glacier and Grand canyons, and fought to protect Rocky Mountain and Mount Rainier parks. As an Interior Department speechwriter, I had quoted John Muir and John Marshall, visionary men who nevertheless had urged depopulation of territories for elite recreation pleasure. My hero, Wallace Stegner, wrote an inspiring "wilderness letter" of a "virgin" untrammeled landscape, a "geography of hope" where man is a visitor but not a resident. My country had expelled Native Americans from a third of its landscapes, like this. "Yes, it has," I finally answered, "even in the not-so-distant past."

Bapalo's gaze did not accuse. Our eyes met. I glanced away.

Then I began to speak in upbeat tones. "That is why your culture and fight are of such powerful interest to Europe and America," I said, to encourage them. "And to me. You still live here. Your refusal to move makes an outcome possible that is now too late for my own country."

Nyare Bapalo broke the silence. In a soft voice he explained, "I had heard from the Botswana officials that America's government had pushed aside the First Peoples in its country. But I did not believe it true until now."

CHAPTER 12

The Paradox of Bling

BUSHMEN COULD NOT COMPETE WITH elephants for the foreign exchange revenues brought by ecotourism, but another luxury linked them to the affluent industrial economy beyond the Kalahari: the lucrative trade in trinkets. On top of her multifaceted career—grocer, tracker butcher, cook, midwife, architect, woodcutter, singer, storyteller, dancer, and pharmacist—it turned out that Qoroxloo was an accomplished jeweler. She mined, processed, graded, sorted, polished, assembled, mounted, and traded Africa's oldest cosmetic adornment, a value-added luxury item that men coveted and women desired: ostrich eggshell beads.

There was an abundant and potentially endless supply of raw material, as Bushmen lured off nesting ostrich, removed a few ovals from their clutches, extracted the protein-rich liquid contents through a hole, and used the shell as a lightweight and leak-proof canteen. When these eventually broke, the raw shards had no inherent value. But Qoroxloo gathered them up and kept them and in her free time broke the jagged bits further into one-square-centimeter pieces.

Twirling an awl between her hands, she drilled a hole in each, a thirty-second task. She strung and rubbed them with a soft stone for twenty minutes, grinding out an even cylinder so that each disk became uniformly round and smooth. From these beads she made necklaces, bracelets, armbands, aprons, headbands, embellishments, or baby harnesses; they shone brightly against reddish skin.

Over time, as the total quantity of beads mounted, the individual shells themselves lost value due to abundance. Their true worth lay in who made them, and how, and the context of exchange. When Qoroxloo gave her work as gifts, recipients knew how much labor went into their manufacture, and the time involved gave them inherent value. An apron containing four thousand beads represented nearly two hundred hours of work, and additional value accrued based on who else had worn the beads and for how long. The goal was to pass gifts along, keep them in circulation, not to hoard them. A recipient of Qoroxloo's necklace treasured it for a while and by accepting it incurred an unspoken but publicly acknowledged debt that would later be appropriately repaid.

Qoroxloo's exchange network went beyond the central Kalahari, as beads worn or carried slipped across borders into the global economy. For the beads, some tourists and businessmen were prepared to exchange items of unheard-of local value, like steak knives, tennis shoes, plastic water containers, and boxes of matches. Some even offered cash, a currency Qoroxloo could neither count nor spend.

Still, anyone reading this book is not likely adorned with ostrich eggshell beads. If you seek out a jeweler and press your face against the glass display cases, imagining a transaction for a special occasion like a marriage proposal or anniversary, the odds are you are looking for something shinier and more durable, perhaps affixed to a band or chain of gold: You are seeking a perfectly cut diamond.

There's no shame or guilt in wanting to invest in luxurious vanity. Qoroxloo appreciated the desire to enhance the beauty of a family member or even herself, to make a public display of love.

And yet a diamond transaction today invites controversy. For starters, once bought a diamond becomes curiously difficult to sell. Also, you can never be quite sure where it came from, or who might have suffered as the raw rock made its way en route to Tiffany's. You must take it as a matter of faith that it has not been artificially manufactured. And even if it is indeed genuine, and half a carat or more, then wherever on earth the diamond is bought the odds are the quiet credit card transaction would trace directly back to Botswana's arid sandscape. There, your desire for a diamond required the destruction of millions of gallons of water from beneath the Kalahari, water that will never be recovered or replenished. It would have led Botswana's government to crush the Bushmen's traditional vanity-item trade economy and make ostrich shell prospecting illegal. Finally, it would have driven officials to cut off water to Qoroxloo's band, scattering her family and besieging her in an effort to empty their ancestral home.

Qoroxloo understood jewelry and adored enhancements to beauty. But her values were such that she could never grasp why so many people were prepared to sacrifice a vital biological necessity in pursuit of a useless industrial luxury. She was in good company. For centuries great thinkers and economists have called this riddle the "diamond-water paradox." In the Kalahari this paradox of misplaced value would obliterate life to extract inanimate old chunks of carbon. The tragedy comes from realizing in hindsight how human nature made such a senseless and destructive outcome all but inevitable.

FROM THE SEAT of tribal royalty Dihabano and Dithunya Mogae sent their son to be educated abroad in England. Festus Mogae attended the North West London Polytechnic, studied development economics at Sussex University, and earned an honors degree in economics from Oxford University, where his economic gospel was Adam Smith's *Inquiry into the Nature and Causes of the Wealth of Nations*. That classic elucidated the beneficial forces driving civilized growth. Human nature's innate "propensity to truck, barter and exchange one thing for another" brought competition, division of labor, efficiencies of the market, and all material goods and services known to humankind. Such insights showed Mogae how he could lift his impoverished country up from its current state of cattle manure and Kalahari sand. The only wrinkle in its logic was what Smith called the *paradox of value*.[1]

By definition, economists measure household values, and their formulas must hold up to reason, not just common sense. And yet "value," Smith fumed, "has two meanings." One came from utility (like Qoroxloo's ostrich egg as protein or water canteen); the other from purchasing power (after she shaped broken shards into jewelry). What tormented Smith was the inverse relationship of these two values: "Nothing is more useful than water: scarce anything can be had in exchange for it. A diamond, on the contrary, has scarce any value in use, but a very great quantity of other goods may frequently be had in exchange for it."[2]

At the time Smith published, in 1776, diamonds were genuinely rare. The annual global supply amounted in size and weight to a single glass of water.

Yet humans could all live without diamonds; we died without that water. So why could a glass of useless diamonds be traded for millions of buckets of vital water? Smith tried to explain away the "diamond-water paradox," noting that it took far more effort to mine, cut, and polish a bucket of diamonds than a bucket of water, so maybe accumulated labor gave luxury items their value.[3] Of course, even Smith conceded that labor didn't explain why an easily found and cut large diamond was worth more than a deeply buried, labor-intensive small diamond, and for the next century the paradox aggravated economists. By the time Festus Mogae landed in England, however, neoclassical economists had derived the obvious solution: $MU = \Delta U/\Delta Q \cong dU/dQ$.[4] I would exchange lots of water for one diamond if I had lots of water and few or no diamonds; you would trade a diamond to get more water only if you had no water and many diamonds. "Diamonds and water have no individual inherent value" was their triumphant conclusion. "They are tradable substitutes for each other!"[5]

This kind of dry logic earned economists their reputation for knowing the price of everything and the value of nothing. The abstract comparison had made sense on paper ever since Copernicus, but as Festus Mogae graduated with honors and returned to his home in the world's fourth-poorest country, the diamond and water paradox seemed obscure and irrelevant.

In reality it wasn't. A century earlier, another Oxford man, returning to impoverished Africa, had founded a global cartel and fueled the ambitions of empire through his original genius to monopolize control over the exchange of millions of buckets of water for millions of buckets of diamonds.

In 1870, told by a London doctor he had four years to live, Cecil John Rhodes sailed to Cape Town, bought prospecting gear, and set off by ox cart to converge on the largest diamond rush in history. On the banks of the Orange River, fifty thousand fortune hunters swarmed like termites around the boomtown of Kimberley. His elder brother Herbert had staked claims on the farm of brothers Nicolaas and Diederick De Beer, but weaker Cecil often shared the thirst of those laboring hard in that hot landscape. Like Levi Strauss during California's gold rush, he realized fortunes could be made not through exploiting Kimberley land, but rather from exploiting its parched miners. Soon Rhodes began selling canteens and jugs of cool, fresh water. Business thrived. Laborers needed more and more water to drink while they dug down, but as mines deepened and widened, a new diamond-water opportunity arose.

What held back mine expansion wasn't capital or labor or land. It was water. So close to the river there was too much groundwater seepage, and owners needed to suck it out. Seizing his chance, Rhodes pooled all his savings, bought the continent's only steam-powered water pump, and sold its services to the highest bidder. With increased cash flow, he bought more pumps and drove out competitors and soon locked in monopoly control of both supply and demand of water. He charged outrageous prices. If owners lacked cash, he was gentlemanly enough to accept generous payment in equity shares. In a decade his water monopoly made Rhodes the largest diamond mine owner on Earth.

Ironically, Rhodes's efficient pumps made mines *too* productive. Just as neoclassical economists predicted, diamonds remained valuable only to the degree they remained scarce. Rhodes's water monopoly taught him the need to self-regulate supply and production—hold back selling until prices went up. "As long as there were autonomous competing mining companies the market would continually be flooded," he warned. "Then prices would fall to the point that the public would realize that diamonds had no intrinsic value." By twisting arms and breaking resistance, Rhodes melded all his independent mines into a single amalgamated entity that operated variously as "the Trading Company," "the Syndicate," "Central Selling Organization," or "Mining Services, Inc." But it was always best known by its birth name, De Beers.

By controlling 95 percent of Earth's diamonds, De Beers ensured supply remained artificially scarce. But it initially had less control over two other scarcities.

The first scarcity was demand. When Depression-era citizens spent incomes on more practical and prosaic items, or patriotic widows sold off useless diamonds to win the war, De Beers feared reversal of demand might crush its inflated prices, and flew into action. It hired N.W. Ayer (now BCom3) to correct such misplaced values with an ingenious marketing campaign. De Beers product placement annually "salted" twenty thousand "news" stories about diamonds in the press. It issued press releases about diamonds worn by Liz Taylor or Elton John, and targeted writers and studio chiefs with story ideas for James Bond novels or Marilyn Monroe movies. Rather than merely sell gemstones, De Beers became the go-between in a quasi-religious transaction in which uncertain young men had the opportunity for just "two months' salary" to endow their betrothed with "a girl's best friend" in a public investment establishing their "status" in love everlasting. De Beers' $200-million annual advertising campaigns proved neoclassical economists wrong: Diamonds were not "substitutes" for commodities like water; they were substitutes for Eternity. Diamonds were Forever. By crystallizing sentiment into facets and persuading couples that reselling diamonds was equal to recycling love, De Beers guaranteed itself an ever-expanding demand. From Festus Mogae's birth to the time he joined its board of directors, De Beers U.S. diamond sales grew from $23 million to $2 billion; when Mogae became president, De Beers monopolized the global market, posted profits 90 percent higher than those of the previous year, and relied on Botswana as its top global producer: the world's biggest source of luxury, gem-quality diamonds.[6]

Boosting human demand was one thing. Yet it was the second, unemotional scarcity that threatened to grind De Beers to a halt.

Water.

Few outside the extractive mineral industry appreciate exactly how much freshwater the entire process inhaled.[7] California's 1848 gold rush lawlessly devoured that state's watersheds and aquifers—particularly in arid Nevada, Colorado, Utah, and Arizona, where the mining industry destroyed twelve thousand miles of America's rivers and 180,000 acres of lakes and reservoirs. In Montana, Phelps Dodge planned to pour cyanide-laced water over ore to extract tiny flecks of gold in a mining process that would lower the water table by 1,300 feet.[8] All told, the world's extractive and processing industries annually suck out 200 trillion gallons of water from the earth's aquifers and streams.[9] Diamond extraction was extremely thirsty, and while it had the cash for desalination technology, Botswana was landlocked far from the ocean. Freshwater was scarce and increasingly expensive. De Beers would consume 11 percent of Botswana's water,[10] but that water was too heavy to be shipped or piped in to remote places like the prodigious mine at Orapa, and the energy costs of pumping groundwater escalated as the wells had to lift more water from greater depths—fossil water that might dry up without warning, gone forever. Visible surface water was cheaper and more reliable, if De Beers could find it.

Its first target for Orapa was the ephemeral Mopipi River. The Mopipi drained the seasonal rains that fell in the Makgadikgadi Pans and, along with the Boteti, fed Lake Xau. De Beers diverted the trickle behind a low retaining wall, essentially privatizing the river to control supply in an artificial reservoir. Yet the fierce sun and wind frequently evaporated that shallow water into concentrated salts. In some years the corporate pond dried up completely, briefly stopping all diamond mining. Casting farther afield for a more reliable permanent source, the cartel set its sights on the Okavango River.

The Okavango gushed out of Angola, ducked across Namibia, channeled down a fifty-mile-long panhandle in Botswana, crossed the Gomare Fault, and then meandered out magically like a many-fingered hand across the Kalahari. The delta of this massive desert terminus river created the world's largest "wetland of international significance." It rose and fell in a flow so delicately balanced that a pod of hippos could quite literally stir ripple effects across the entire ecosystem. De Beers planned to dredge, straighten, sludge-transfer, and line twenty miles of one of the Okavango's fingers, cutting deep into the delta.

The plan was ambitious. It would speed up water flows; store them in shallow dams along river valleys; supply water for the safari town of Maun; and cut through the fault line in order to release water flowing down the Boro Channel, out through the Thamalakane and finally the Boteti River to its desert destination near Orapa. Hydrological engineers saw a challenge. Development economists saw growth. De Beers saw lower production costs. Foreigners saw unmitigated catastrophe and mounted an aggressive campaign to halt the scheme.

Setting a precedent, the high-profile "Diamonds are for Death" protests led De Beers to ostentatiously withdraw, while quietly encouraging the government to proceed. Meanwhile, men and women from local tribes vowed to take up arms in resistance to the project, claiming that "God made rivers meander for a purpose." All impressed upon the government their vehement opposition. Following a flurry of cabinet-level meetings—including the then minister of finance and development planning Festus Mogae—the government hit the pause button. It requested a study from the World Conservation Union (IUCN), whose scientists concluded the project was fatally flawed, harming six of the delta's nine zones. Botswana's president Quett Masire suspended the project indefinitely, forcing diamond mining to live within the limits imposed by water. Never again would the country and its industrial benefactor be so transparent about its water development plans.

A decade later, when Botswana officials told Qoroxloo and one thousand other Kalahari Bushmen that they must leave, cut off their water, and laid siege to their reserve, very few managed to explain why. Minister of Local Government Nasha shrugged, noting the precedent of those previously transplanted in Jwaneng "to give way for projects of national interest," which in that case happened to be a diamond mine. Foreign Minister Mompati Merafhe added that he, too, was perplexed by the international uproar. "Many Bushmen have been removed because of economic interests," he explained. "In Orapa, my area, a great chunk of people were removed because of the mine. Botswana is where it is today because of this facilitation. These people are no exception."

Yet the uproar continued and as the forced evictions triggered a new round of international boycotts, the "obvious" explanation got shushed and new rationales cropped up. Officials now asserted in public and in court that Bushmen relocation was for "development" or "health" or "wildlife." De Beers maintained that the relocation had nothing to do with its diamonds. Dr. Akolang R. Tombale, permanent secretary of the Ministry of Minerals, Energy, and Water Resources, affirmed that the evictions had nothing whatsoever to do with diamonds. Even Botswana's human rights groups reiterated that cutting off water had "nothing to do with diamonds."

Indeed, in the eyes of Botswana apologists, De Beers was just an innocent but fat, sexy, glittery target for Bushmen publicity hounds seeking to plunder De Beers' deep pockets and to exploit its consumer-fearing Achilles' heel. Besides, by soiling the company's reputation, dissident Bushmen were strangling the goose that laid the diamond egg, the source of revenues that catapulted the country's per capita annual income from $80 at independence to $5,900 today.

Once again De Beers publicly distanced itself from controversy while doing nothing to change government policy. It hired a respected anthropologist to argue on its behalf as being Bushmen friendly, and claimed any diamonds found inside the reserve could be extracted without touching a single grass hut. In any case, De Beers maintained, these "inflammatory campaigns" and "unjustified accusations" by Bushmen and their allies were economically moot. "Constraints in the market" revealed "little demand," so that the Kalahari's potential diamond mines were "dormant" and considered "uneconomic." But as De Beers knew better than most, value was relative to scarcity, and a "diamond-water paradox" that Smith hadn't foreseen was that only the former would be worth massive artificial replication.

SINCE THE DAWN of existence, water was a phenomenon exclusively generated by nature. Water's manufacture required force beyond human capacity. It could not be created or synthesized. "Water," said tribal Africans, "was a gift from God."

De Beers' chairman Ernest Oppenheimer felt the same way about his gemstone. Pure carbon, or graphite, had to be squeezed at high temperatures and two hundred thousand times atmospheric pressure. At the moment of their bursting up through the magma, Earth's energy compressed carbon sheets under Herculean heat into hexagonal rings and transformed them into carbon atoms linked in tetrahedral bonds (adding considerable value en route). "Only God can make a diamond."

In both cases, God has worked in mysterious ways.

Henry Cavendish liberated a light gas that burned readily in air; Joseph Priestley "discovered" a candle-fueling gas that helped his own breathing. After igniting and exploding these isolated "impure airs," the scientists noted how instead of smoke or soot, the accumulated stuff on their tools was moisture and dew. The Frenchman Antoine-Laurent Lavoisier named the pure air *oxygène,* or acid former; and the flammable, dew-producing air *hydrogène,* or water former. After forcibly combining both elements to synthesize water, Lavoisier then forcibly split it back into the two elements, breaking water apart. Later Enlightenment thinkers proved it had become possible to construct water.[11]

Scientists later proved Oppenheimer wrong concerning Who could transform carbon. Within decades General Electric's physicists managed to force one million pounds per square inch within tungsten carbide walls that withstood five-thousand-degree heat. By 1955 they transformed tiny, genuine, low-grade crystals: manufactured diamonds.

De Beers' stock plummeted at the news. Suddenly it faced competition. Like Rhodes at Kimberley, the cartel bought up seventy-five of the hydraulic presses. It harnessed water to pump out diamonds in the same land where it once pumped out water to harness diamonds. True, the product was low-grade industrial "boart" diamonds, but they represented a fourth of the market and cost less to forge than to mine "real" diamonds. The rules of the game were changing fast. De Beers rushed to open the mine at Orapa just as GE pressed its first gem-quality diamond larger than a carat. There was enough real wet stuff out there to negate any value in synthetic water. Likewise, economics ruled against GE's plunging into De Beers' saturated diamond market. Stamping out more diamonds only made them cheaper. Still, De Beers realized it was now in a race against time: In 1980 nine out of ten industrial-grade diamonds were synthetic; a decade later De Beers reportedly made gem-quality diamonds along with mined diamonds; and by 2000 no experts could tell the difference. To dealers, this was the equivalent of counterfeiting one-hundred-dollar bills that could never be detected. The global market could evaporate Botswana's entire economy. To avert that fate, De Beers ingeniously "branded" its mined stones with a microscopic "Forevermark" inscription invisible to consumers yet requiring their complete trust in the company's word of honor. Reversing a century of playing up rarity as being the basis for a diamond's worth, the market leader in a $62-billion industry resorted to the now-discredited Marxist theorizing that a product's inherent value came from the many hours of costly labor that accumulated in the process of making it.[12]

In 1998 Harry Oppenheimer gave De Beers to his son Nicky, and Quett Masire handed Botswana to his chosen successor, Festus Mogae. The next year, De Beers unloaded half its accumulated diamond stockpile, flooding the global market. Within two years De Beers bought up its public shares to transform itself into a private company 90 percent controlled by the Oppenheimer family. The company's remaining 10 percent belonged to the Debswana joint venture, half owned by Botswana, and controlled by De Beers. Going private was expensive but allowed more secrecy and freedom from shareholder interference. On the advice of Bain & Company, De Beers crystallized its name into a brand. It would compete on its own, with a narrower but still vertically integrated system from the pit mines in Botswana, through its cutting and sorting process, to launching what would become its first elegant jewelry outlet in London, with David Bowie's supermodel wife, Iman, scheduled to highlight a celebrity-only, diamond-studded evening.

It didn't turn out quite as planned. Bushmen protesters upstaged the event, co-opting De Beers' own marketing slogan while turning the new brand's coming-out party into a referendum on Botswana. It substituted the image of a hot young supermodel sporting a diamond with a wrinkled Kalahari elder adorned with ostrich eggshell beads under the slogan: BUSHMEN AREN'T FOREVER.

Nor was De Beers' monopoly. As the cartel's grip weakened, Botswana's firmed up. The country now produced between a quarter and a third of the global market for rough stones and had other international suitors. President Mogae wanted to do more than just supply. Like Qoroxloo with her eggshell beads, he wanted to mine, process, sift, sort, aggregate, sell, cut, and polish raw material, here at home, adding 50 percent to the value of the stones. Unlike Qoroxloo, he got what he wanted. By 2009 all De Beers stones from around the world were to be processed and refined in a new building under construction in Botswana's capital. In exchange, De Beers got an exclusive commitment from the company's main source of production. "Botswana is the jewel in the De Beers crown," explained the journalist Janine Roberts, who covered the cartel for fifteen years. "It is the country, above all others, whose diamond production De Beers must control if it is to maintain its stranglehold over the diamond market."

For 130 years, Rhodes's monopoly affirmed "the only way to increase the value of diamonds is to make them scarce, that is to reduce production."[13] Now, open competition reversed priorities from hoarding its stockpile to accelerating output. Dormant, formerly uneconomic and low yield mines had to be pushed into production. One such mine in the central Kalahari lay at the heart of an ancient Bushmen hunting, gathering, and burial ground, a place known as Gope, meaning Nowhere. Another reportedly sat beneath Qoroxloo's grass hut at Metsiamenong. Labor and capital costs didn't make these potential mines uneconomic and were not holding back production.[14] But for the time being, lack of water was.

A boilerplate environmental impact statement projected that the Gope mine would require a billion gallons of water per year, for decades. That water volume eclipsed the health needs of the country's entire population, but since average water consumption of mines was 170 gallons per carat, it revealed that Gope would annually produce about 6.5 million carats, valued at half a billion dollars, the world's fifth-largest diamond mine.[15]

So what will it be: a billion gallons of water or a half billion dollars of diamonds? Botswana's choice would be obvious, and its consequences irreversible. Every drop of fossil water that De Beers used would be gone forever. Aggressive pumping by just a single mine would suck down the water table in a radius spreading up to hundreds of miles in all directions, wreaking havoc on entire populations, just as it had in America's arid West.[16]

Therein lay the final paradox. On Shoshone ancestral lands Nevada extracted $21 billion of minerals, while on Bushmen ancestral lands Debswana extracted billions more in diamonds. Industries and consumers gave minerals a defined price tag, yet all considered water to be valueless. Qoroxloo found that mining threatened the end of all that she valued in life, for water to her was priceless.

In our larger industrialized world, far outside the Kalahari reserve, government subsidies have encouraged luxury goods manufacturers to burn up trillions of gallons of water in an increasingly hot and arid landscape. None of us have any real idea what all that water might actually have been worth. We don't know how much of the true foundation of the wealth of nations has been sacrificed forever. It was only a matter of time before we would find out.

PART IV

Constriction

CHAPTER 13

Oriented Against the Sun

BUSHMEN FELT IT PAINFULLY OBVIOUS THAT Botswana was trying to evict them for the sake of more lucrative cattle ranching, ecotourism, and diamond mining. But President Mogae denied it. He maintained to foreign ambassadors that his government cut off Kalahari water deliveries simply to improve and uplift the wretched lives of these Stone Age creatures. Besides, he added, Botswana was a sovereign state and could do as it pleased. His taxpayers had fully compensated any displaced individuals for their troubles, and the state owed Bushmen nothing further.

On that matter Mogae was at least technically correct. His officials had meticulously measured, described, and listed anything people categorically defined as property, and then provided cash compensation to offset the loss of assets from a Bushman's surrendered home. Even the Bushmen's attorney conceded the process had been diligent, honest, and perhaps even generous. President Mogae had been as fair as any insurance actuary in his calculations, and as such he entirely missed the point: Certain losses can never be replaced.

To Qoroxloo and other defiant Bushmen, the assets locked up in the concept of *home* meant something more than a spreadsheet tabulation of the square feet of grass-covered dome huts, the scavenged bags and blankets, the ostrich eggshell canteens and plastic containers, a dog, goat, three chickens, all contained in a yard plot demarcated by sticks in the sand and a melon patch. Home was more than a loosely assembled collection of tangible objects. It was the interwoven habits, the smell of woodsmoke blown in a known direction, or the tart taste of wild monkey orange fruit growing impossibly in the desert on the same tree year after year. Home was the sound of a mouth harp or thumb piano followed by ancestral myths recounted slowly with long pauses to people who knew them by heart and enjoyed correcting the storyteller if he left out a seemingly inconsequential detail about the place where the event took place. Home was the teasing and joking relationship that bound a grandson to Qoroxloo and bound her to deceased ancestors who were never really gone because when called they answered, providing roots and security in the vast expanse of desert and sky. These subjective associations could not be quantified. But they reminded Qoroxloo's band where and who they were, and became more powerful than anything that could be lifted onto the bed of a truck and hauled away. The ties grew from the place itself, which could not, by definition, be physically removed or liquidated into cash. And because our species values most that which is rare and scarce, the most precious assets of home, for Qoroxloo, meant water.

In that arid land most of the Bushmen's acts and decisions—activity, dispersal, sanitation, gathering, hunting, strategies, luxuries, and politics—revolved around the need for and access to water. Water resources gave shape and definition to circular or elliptical units of land. These units were recognized both at the individual and group levels, as a home territory, or to Gana in the reserve, a *g!u*.

The physical borders of an unmarked *g!u* cannot be found on paper or in court records filed as a title deed. Yet Bushmen like Qoroxloo bound their identity to it, recognized its extent and limits, and restricted access and use of its limited resources, especially in times of drought. In each Bushman *g!u*, access to the Kalahari's precious water resource base was not open to all and blindly egalitarian, but rather individually determined on the basis of the water resource base, how many planned to use it, and for what strategies.[1]

Once again, by organizing the use and interaction of territorial and property resources, water governed people.

Water-governed territories were usually bequeathed along kinship lines in an unwritten title transfer that was reinforced by songs and stories that blurred the lines between myth and reality. One such story describes the water-based founding of Qoroxloo's home. Long before those strangers known as Black People or White People ever even visited the Kalahari land still inhabited only by the Red People, or Bushmen, a young man by the name of Sa'ama went out hunting alone and walked far from where his band had camped. Sa'ama opened his senses in order to read whatever signs he could, and on that day at a great distance he saw open-winged vultures wheeling dark against the blue sky. One by one the vultures descended to the earth and landed at a vanishing point on the horizon. He knew these hooded, white-backed or lappet-faced carrion eaters, and if he chased such scavengers from a carcass, the remainder would be his. Sa'ama hurried ahead over the sands to discover how much flesh might still be clinging to bones. Drawing near, he noticed no cloying smell of blood and putrefaction, but as he entered the clearing, his doubts turned to ecstasy.

The vultures huddled in a circle were not eating; they were drinking.

A puddle no larger than an eland hide had miraculously pooled on the sand. Only a compressed thin layer of nonporous mineral, called calcrete, prevented gravity from pulling the water down through the sands and out of reach. Sa'ama chased off the vultures and drank deeply. Knowing its worth, he hurried back to the clan and told one of the elders, Mozampo, of this water. After brief discussion, the band decided to converge on that area, thereafter known as Qxhaetsha, or "vulture pan." The name stuck but was much later translated by the Black People into metsi (water) a (of) menong (vulture), or Metsiamenong.

Qoroxloo had inherited the small pan, passed down through the generations from her ancestors, and through intricate informal arrangements more binding than any contract the members of her band shared in the task of guarding and organizing their lives and activities largely around its water. Because it represented their home's most treasured asset and namesake, and because it had been bequeathed uniquely down through time to the current inhabitants, Qoroxloo and her ancestors improved the tiny basin over time through careful excavation. They made it deeper and easier to scoop from the bottom and kept steel drums and plastic containers ever at the ready to secure drops as soon as they fell, protecting their captive waters from evaporation. People came and went with the seasons. But Metsiamenong remained, linking generations as the overriding orientation.

Even so it was but one water resource that defined Bushmen territories, for its primary utility arose during the brief, sudden rainy season when pans gathered and filled quickly. But they could dry up even faster.

It is no surprise that Bushmen stories rank water and land among the "life things" that give, while the sun, like the hyena, is a "death thing" that takes away.[2] Against that ancient, ruthless enemy the Kalahari Bushmen were locked in a constant struggle, forever struggling to secure, lock up, and withhold water from the sun.

THE MODERN NATION-STATE of Botswana had its own creation story, which also revolved around water. Yet instead of chasing vultures, the country's leaders had chased after foreign development aid; instead of advising it to conceal water from the sun, the overseas experts assured Botswana it had nothing to hide: Engineers explained that a brand-new, big, strong, and centralized state deserved to rest on brand-new, big, strong, and centralized dams.

True, dams were not new technology; water storage dams of Jordan, Egypt, and other parts of the Middle East dated back three millennia before Christ. But during the twentieth century that basic concept was put on steroids, peaking just as Botswana was born. Over several decades one hundred nations spent $2 trillion to build forty-nine thousand dams larger than a four-story building.[3] Through the miracle of large dams, the once-impoverished Norway had electrified and enriched its sleepy fjords; formerly barren Australia had impounded 5 trillion gallons into lucrative agricultural exports, and Depression-afflicted America had tamed and transformed great rivers from the Columbia basin to the Tennessee Valley into economic powerhouses. The New Deal's cornerstone, Hoover Dam, plugged the Colorado River behind a smooth concave wall consisting of 66 million tons of concrete and rising 726 feet, backing up 9 trillion gallons of freshwater—enough to cover Connecticut ten feet deep—into the 115-mile-long Lake Mead, at that time the biggest, deepest artificial lake on Earth. To dedicate it, on September 30, 1935, President Franklin D. Roosevelt proclaimed of the dam's iconic power: "I came, I saw, and I was conquered."[4]

For any sovereign nation-state government, the conquest and control of water resources is a paramount concern. So from its birth, and increasingly over the next four decades of growth in an arid land, Botswana claimed monopoly power over all rain that fell, all runoff, all aquifers, in short all water that everyone needed in order to live. The future of its modern cities and complex economy would be literally cemented in place, inextricably anchored to a seemingly secure technological foundation. Those present at the creation of Botswana's capital city thrilled as "rising out of the virgin Bush at Gaberones . . . an earth wall nearly a mile wide dammed the Notwani riverbed and after a few heavy showers of rain, a dam nearly seven miles long was created. Long before the new town was built its water supply was ensured."[5]

Or so it seemed. Even in America, as Hoover Dam's concrete began to set, a few critics began to raise doubts about the hidden costs that were submerged by silent, rising waters. Some questioned the logic of pouring water on a desert until the landscape became a waterlogged and saline wasteland, or of destroying robust protein-rich fisheries to subsidize biofuel crops, or of concentrating reservoirs in unstable mountain canyons until the sheer mass of dense water combined with lubrication of tectonic plates to trigger manmade earthquakes that cracked the dam wall and risked populations living downstream.[6] More recently, NASA geophysicists noted how dams' rapid consolidation of streams near the equator, combined with the weight of water collected in reservoirs, have slightly tilted our planet's axis and altered the shape of its gravitational field. To quench our thirst, it appears we may also have distorted Earth's orbital rotation.[7]

Most dam critics have been silenced by crushing demands from a demographically exploding world. Agencies and development boosters counter that as urban populations continue to rocket each year, more and more government dams have had to send 1.3 trillion gallons of freshwater[8] racing through pipes into billions of toilets, showers, sinks, garden hoses, and pipes to cities for domestic, industrial, and commercial use. The best way for big cities to harness that water, they continue to argue, has to be a dam. So Botswana's citizens weren't alone in their dependence on a single monolithic artificial reservoir. The security and stability within most arid nations, including the United States, rose and fell in direct proportion to the level of water trapped behind dam walls.

Poised on the edge of the Kalahari, the Gaborone Dam appeared like a mirage, too good to be real. Most of the reservoir stood less than ten feet deep. It gathered an average twenty-one inches of raindrops that fell each year within a catchment area a fifth the size of New Jersey. Engineers initially designed it to hold 37 billion gallons, which boiled down to fifty gallons per person per day. That left little margin for error. Yet just as Hoover and Glen Canyon dams supplied booming western cities like Phoenix, Las Vegas, and Los Angeles, so Gaborone Dam supported the fastest-growing city in Africa.

With that water Botswana's president Mogae could realize a politician's dreams. He planned to generate jobs and attract foreign investment to make his capital city a continental financial center, a knowledge hub, a technology economy, which could soon overtake Johannesburg and dominate the continent. At current growth and graduation rates, his ambitions appeared well within reach and his vision anchored in reality. After all, Mogae managed his economy very carefully, and, at least in the beginning of his term Gaborone Dam was still full.

Across southern Africa, however, such a state of watery abundance was fleeting.[9] From the day of the dam's birth the subcontinent's brief wet season had declined another fifth while aridity cut stream runoff in half.[10] Compounding the loss of rain, evaporation sucked the moisture content out of thin, sandy soils across semi-arid savanna grasslands.[11] Ominously if predictably, the waters of Gaborone Dam began evaporating, inch by precious inch.

Botswana's Water Utilities Corporation closely tracked, measured, and extrapolated from water levels against the dam wall. From the start of the siege over the course of 2002, the full dam had dipped to 79 percent of capacity; by late 2003 it hit 68 percent, dropping in 2004 to 54 percent. Fifty gallons per person had shrunk to twenty-five. Designated workers warily reported the reservoir's status to the Ministry of Water, whose officials gave the sobering news an upbeat spin. The reservoir had gone down in the past, they conceded, and while it had never sunk so far, so fast, contraction was inevitable in landscapes with variable rainfall patterns that were largely seen as cyclical. This was the very point of dams: to save water in wet years to have around in dry years.

That safe assumption failed to absorb the volatile dynamics of thirst, in which affluence compounded urban growth. The rich use exponentially more water than the poor; ten gallons for survival becomes a thousand gallons for comfort. Gaborone Dam was built to serve a small, sleepy capital village, but within four decades the capital boasted nine shopping malls, a national brewery, a dozen office parks, and two hundred thousand thirsty, sprawling people hiring cheap labor to scrub swimming pools, hose down driveways, water exotic gardens, wash air-conditioned cars, and help them forget that they lived on the edge of an encroaching desert.

Ominously, the "wet years" had begun to taper off. For four decades the mean annual rainfall grew ever more feeble, random, and scarce even as the mercury notched up a degree. Engineers raised the dam wall to hold an extra year's storage capacity, but they could not fill empty riverbeds or store rain that refused to fall.

As frontier towns like Los Angeles sprawled into megalopolises of ten million, America shared Botswana's compound pain of booming populations and shrinking freshwater.[12] The heat was excruciating. Each year the two biggest Colorado River reservoirs evaporated enough fresh drinking water to slake the thirst of every human on Earth for seven months.[13] Meanwhile, peer-reviewed reports project thirteen years ahead to where Lake Powell becomes an inland delta as Lake Mead dries up completely.[14]

While dry heat and wind inhale a seventh of the water above, layers of sediment accumulate below, eating up reservoirs at one percent per year.[15] When the Kalahari water siege began, two fifths of Gaborone Dam's original water storage capacity was already filled with mud. "In any hot arid place like Africa, building a dam is like offering a sacrifice to the gods," the regional hydrologist Christine Colvin explained to me. "All those open canals and exposed reservoirs were like putting giant cast-iron pans out in the desert and saying: 'Come and get it!' "

Climate change increased erosion, heat, and wind, and all three compound forces conspired against dams like Gaborone's. Within months of cutting off waters to Qoroxloo and her band of Bushmen, the government faced the potential end of its own supply.

To keep its reservoirs from floating away, Botswana needed to put a lid on them. But on vast shallow dams in hot dry places, stopping evaporation was easier said than done. First engineers had tried pouring oil on top of the water, to create a film barrier. Later they weighed spraying liquid silicone coatings: thin, environmentally friendly, protective layers. In yet another approach they unrolled plastic sheeting, a sort of Glad Wrap, over their giant urban broth.[16] The good news: These experiments reduce evaporation 40 percent in small-scale tests. The bad news: Outside the laboratory, the sun's ultraviolet rays dissolved chemicals while strong winds blew sheets away overnight. Botswana eventually gave up.

A more promising tactic, arriving too late, aimed to prevent exposure through *artificial aquifer recharge*. Though the tools and scale were new, the concept was as old as the Bushmen. Just as Qoroxloo transferred, sealed, and buried water in ostrich eggs, plastic containers, or clean barrels as a future, evaporation-proof reserve against the sun, artificial recharge forced water down into the aquifer fissures and pores beneath cities. Every ten gallons of water pushed underground saved four gallons from evaporation, allowing flexibility and security. The tactic did not make dams obsolete. To the contrary, it enhanced their function as collectors, and, like the Bushmen's use of the pans at Metsiamenong, dams became merely a means of more efficient decentralized storage elsewhere. As the groundwater expert Dr. Alan Simmonds put it to me, "Once you've grabbed the slippery beast you've got to get it underground as fast as possible before it can escape."

Western experts who once instructed Botswana officials to build large dams on thin rivers and keep them full for as long as possible now told them to rapidly empty the dams to double their water storage capacity.[17] It seemed madness, yet drought and evaporation left no alternative. So Botswana embraced these technologically sophisticated hydrological tactics, which Qoroxloo's band had perfected over the eons.

Unfortunately, an analysis of the best potential aquifer found it contaminated with nitrate pollution, making artificial recharge difficult, if not impossible.[18] As the level of Gaborone's reservoir continued to plummet,[19] Botswana realized it had done what no Bushmen, or any species, ever would do: sacrifice and foul the very waters it would need to survive.

Botswana was belatedly discovering how certain things, once gone, could never be replaced.

CHAPTER 14

Cradling Every Drop

THE BUSHMEN STRATEGY WAS NOT ONLY to secure water away from the sun, but never lose it in transport and delivery to the body. Qoroxloo taught her grandchildren how to care for that vital water as if it were one of them. For indeed it was—more than any other substance, water carried the life force.

In that spirit, people carried the water. Bushmen continued relentlessly to forage and scavenge water, checking tree pits for small puddles that the branches or trunk might have trapped and stored in the shade. Depressions in the trunks of baobab, mongongo, marula, or Shepherd's trees in particular collected rain that could be sipped through a straw made from a wild dagga plant, or soaked up with leaves and then sucked dry. During one brief downpour I watched as Bushmen rushed to attach closed containers onto trunks or lower tree branches, diverting trickles into a small opening at the top, harvesting rain like a fast-forward vision of Vermont farmers harvesting sap for maple syrup.

Beyond tapping these open pans, natural runoff, and exposed water pockets, some Bushmen territories oriented around sip-wells.

The subterranean sip-well is perhaps the oldest form of pumped groundwater on Earth. It can't be dug just anywhere. Most of the Kalahari was pure deep sand, but over the centuries, in a few random places, rain picked up, bonded, and then shaped calcrete minerals into a thin, bowl-like, layered membrane. Rainwater captured between that membrane and windblown sand was protected from sun above and gravity below. Scattered and invisible from the sky, sip-wells were the Kalahari's best-kept secret. Fortunate Bushmen could return to them again and again to drink with confidence, but unless shared with the next generation, the practice would become—like Japanese Zen calligraphy or New England dry masonry—a lost art.

I discovered the extent to which sip-wells demanded deep respect and restraint. Before tapping one, Bushmen sought a blessing from God (or gods, ancestors, or all the above) to guide their scoops, sticks, grasses, strings, and a special herb that was supposed to bring aid to the digging, when smoked. I watched as Bushmen danced around the sip-well in a kind of wild figure-eight pattern, clapping, brushing against each other, chanting a song or singing a chant until they dug down several feet and sipped up the trapped water through a natural straw tube.[1]

A similar level of intensity governed water's transport. Qoroxloo's band could trace every drop of their water directly to the sky, and so did not have to worry whether it had been contaminated at some point along the way as it was conveyed by mysterious transport.

Once again, nature pointed their way. In mornings and evenings, Kalahari sand grouse fly by the thousands, traveling up to forty miles from home to the nearest source of water. The adult male has tightly coiled belly feathers; the filaments uncoil in contact with the water, trapping water like a sponge. In flight the feathers cling close to his body. When he returns to the nest, his chicks then use their beaks to strip water from his wet chest, and drink.[2]

Qoroxloo's band could not match that near-perfect water efficiency, but it came pretty close. Bushmen children grew up learning to handle water with similar care. Until aridity itself could imprint on young adults the value of water, parents applied the severe daily lessons to their offspring. However tolerant and open-spirited, Bushmen restrictions nevertheless grew sharp in the seasons of scarcity and the days of drought. Long renowned for gentle behavior, Bushmen grew visibly angry, censuring each other over any careless or excessive use of water. Children were publicly rebuked into shame if they dropped or sloshed precious water. Even accidental spills got punished. Reticent and shy, Bushmen shouted out angry warnings to a visiting white American camper when I was, stupidly, about to toss out some dirty dishwater. It could all be reused, somehow, somewhere.

To cut losses, whenever possible, Bushmen reduced distances of water transport. They moved closer to where the water was contained, using it on site. But invariably Bushmen also had to carry water from source to camp to mouth. Usually they covered the distance on foot over miles; a few times during the siege they borrowed vehicles of Bushmen outside the reserve to hide smuggled water in caches concealed from the government. In the past, water was stored and carried in hollowed ostrich eggshells and elastic bladders of springhares. Now people also used more durable tin cans, iron pots, plastic bottles, trash bags, and steel drum barrels. Bushmen regularly tested containers to ensure they did not leak. And they kept track of quantities. Somehow, everyone always knew how much or how little water remained, and adjusted demands accordingly.

To help a layperson like myself appreciate the full extent to which water governed the psyche of human lives in the Kalahari, the renowned anthropologist Robert Hitchcock had me consider the nuances embedded in their language. Reports may or may not be true that Inuit people of the Arctic have three dozen different names for ice or snow, but Kalahari researchers[3] documented how various bands of Bushmen used the following words for *water* and the term *to suck*:

Water: /qhaa
Water: g!u, dohmsoan (ka)
Water Acacia: /oa'//g!u!'an (ka)
Water bag: /guug!u/haian (ha)
Water course: /nau, /num
Water-storing tree: oggo
Water that has collected in the hollow of a tree: g!u!'an (ka)
To water a garden: tczq
To make waterproof: thobo
Waterhole: g!u-n!ang (ka)
Waterhole connected to a pan by a furrow: g//ae (ha)
Water in one's eyes: tcaq
Water bottle: g!utan (ka)
Well: tsaho
Well (pit): !'han
Sip: kx'aha; dyshxo'la (avoiding dregs)
Sip (hot liquid): sam or tcam
Sip (cold liquid): qom

Suck: //'ube
Suck: qum
Suck and dissolve: /oy/i
Suck (slurp moisture through a small puncture): /qhulu
Suck out: //oho
Suck out blood through a straw: goo
Suck out water and transfer to a container: txhxole
Suck out water from an egg or water tree: Ooumi

This list suggests the numerous ways and the careful extent to which Bushmen restrained themselves; restricted access; wasted nothing; sought out, hunted down, and trapped water as soon as it formed. Qoroxloo's band organized around small, secret, and decentralized reservoirs, sealed off not only from the hot sun above but also from the thirsty sand below. In the process Bushmen revealed the entire Kalahari as the most efficient form of storage and conveyance, interacting with, and depending on, what might be called their own natural water infrastructure.

NATURE WAS OLD, slow, and required local knowledge and autonomy. Botswana was modern, speedy, huge, and vertically integrated. So the young government, like all nation-states, preferred to build extensive new artificial water infrastructure, laying pipes as fast as it could to distribute as much as it could to as many as it could for as little cost as it could. Sure, some spilled along the way, but that seemed a small price to pay for fast growth, consolidation, and nation building. How much was sacrificed in transport? Frankly, the answer to that question didn't particularly interest most government officials. The only do-gooder who seemed to care was soon isolated as a bit of an annoyance.

Chris Schaan was an ex–Peace Corps volunteer in Botswana who somehow retained his youthful American ideals. Raised in Nevada, he recognized how drought was a perpetual problem in his adopted country, too. So he took a job with the Department of Water Affairs, hoping to help it alleviate the country's constant thirst, and traced the watery lifelines from source to point of consumption, documenting what he found, starting at the department's headquarters.

Nowhere did he encounter either vigilance or restraint. In the middle of a drought, top Water Affairs officials washed their cars in the parking lot, letting the hose blast across steaming asphalt and drain away. The headquarters' own cracked pipes bled three gallons of water into the sand for every gallon used. The government lavished a misguided $7 million to install shiny new pumps, which just forced higher pressure through a broken system so that water losses soared even higher.

Leaks were systemic and ubiquitous throughout the nation. Each day, through bad pipes, malfunctioning toilets, neglected hand basins, and auto-flushing urinals, Botswana's prisons sacrificed three thousand gallons; its Defense Force gushed eight thousand gallons; its university lost half its water. Schaan soon discovered a perverse dynamic: Buildings bled far more water when people were away or asleep than during hours of peak consumption. Why? Because as pressure built up during the night more water was forced out through cracks, sometimes doubling or tripling the leakage; during holidays, one school leaked nine times what it used while in session, and, taken together, the country's one thousand schools were on average wasting three hundred gallons per hour.

Most cracked mains and burst valves were concealed and ignored; people drove over pipes left in the road. In Serowe, a four-inch-diameter pipe gushed like a fountain while nearby, young vandals played soccer with a plastic ball from its pressure valve. Squatters illegally tapped into water mains. Going from village to village, Schaan found hundreds of leaks everywhere he turned, "soaking into the sand, seeping through fractured rock, artificial marshes that resembled Australian billabongs. Chambers full of potable water and green acres and wetlands where they should not be."

What infuriated Schaan more than the leaks was the utter lack of accountability. He shared what he found with top government officials, even the vice president, expecting them to rectify the problem: trace the leaks, plug them fast, and fine or at least scold the offender. Nothing. The entire system ran on perverse incentives. Charged tens of millions each year for water that bled through leaky pipes, Botswana's schools attached their water bills to a federal form to be paid in full by the Ministry of Finance.[4] Thus subsidized, the Water Utilities Corporation annually earned profits of $39 million, handsomely rewarded in a drought year for wasting more water than ever. "Its biggest client is the government," said Schaan, "which is also its biggest waster."

Botswana was hardly unique.[5] Research from Los Angeles to New Delhi to São Paolo and Mexico City found that in nearly all big cities, 40 to 60 percent of water never reached the consumer due to leaks, corrosion, disincentives, and poor maintenance; farmers lost two thirds of their water for the same reason.[6] India's pipes sacrificed half the nation's water between source and delivery point, yet farmers forbade even the installation of meters. China, the earth's thirstiest country, wasted more water from aging infrastructure than it could store. Western democracies were scarcely immune;[7] even during drought years the United States leaked 6 billion gallons of water each day.[8]

Still, water remained the state's resource to use or abuse as federal officials saw fit. Water infrastructure is never really considered a serious political issue, economic risk, or national security threat. At least not until the nation in question has almost nothing left to leak.

CHAPTER 15

The Reckoning

BY LATE 2004, AFTER A DRY AUSTRAL WINTER, Botswana's primary dam stood at 27 percent of capacity—the lowest in its history. Experts gave it seven months. Nervous officials tried to assure the public by announcing that "in the event that the Gaborone Dam runs dry," the government "has come up with strategies to deal with the situation."

Botswana initially failed to specify what kind of strategies it had in mind. But months later, one hot day as Gaborone Dam dipped further to another record low, water officials fanned out across the capital city. They rang bells, shoved pieces of paper through doors and gates and mail slots, and departed. The text was uniform, printed on tens of thousands of copies, and reached every home, large or small, to break the news: With twelve gallons per citizen under storage, mandatory austerity was in order. No watering, sprinkling, irrigation, or hosing down pavement. No water on playing fields, parks, gardens, or cars. No urinal flushing. No water in swimming pools. No more cheap water. After that, Botswana rationed what little trickled out. Unable to block the sun, unwilling to plug its leaks, the government forced citizens to swallow a pill that was arguably more bitter, destructive, and odious than taxes. Just as it had first imposed extreme limits on the Bushmen, Botswana would now restrict and perhaps cut off freshwater delivery to every citizen.

But was restriction the solution? Worldwide, cities had brought similar rations, from Los Angeles in the 1970s to Atlanta in 2007. While such draconian measures worked, at least temporarily—within six months Botswana reported a 35 percent drop in water use[1]—they often exacted a heavy toll. In the process of rationing water, the governments wrought sudden, inequitable, and often permanent devastation. Trust was shaken. Water-intensive companies were forced out of business. Real estate prices plunged, undermining new development and economic growth. Ruling parties lost support. Even water utilities risked slitting their own throats. Rationing reduced their operating revenues and budget; rather than burning off fat, the headquarters office was predisposed to lay off its most skilled and experienced workers, leaving only the slowest, least-motivated, and poorest-paid staff less able to monitor dam levels, investigate leaky pipes, explore coping strategies, or enforce the rationing it had just undertaken.[2] Rather than austerity, water rationing often bred incompetence. Water restriction not only curtailed civic freedom; it also shrank the ability of governments to rule.

Week after week the restrictions clamped down on Botswana. Lawns browned while exotic flowers and shrubs drooped and cracked. In early 2005, the Water Utilities Corporation spokeswoman enthused that Gaborone Dam was still 23 percent full (as opposed to 77 percent empty) but warned that "those who did not comply" with rations and restrictions "would face the full wrath of the law."[3] She wasn't joking. The government set up a hotline for snitches to inform on neighbors, who were slapped with fines and penalties.

Extreme reactive measures become the default option of water-stressed countries. Even in the liberty-loving United States, drought-struck southeastern counties won a broad public mandate to unleash nighttime patrols that cracked down on unauthorized outdoor watering and shut off the taps of repeat offenders. As they faced stiff fines and high prices, Americans turned on each other. "We're taking hundreds of calls from people who are ratting out their neighbors," said Janet Ward of Atlanta's Department of Watershed Management. "People are angry. What do you expect? If your lawn is brown and blowing away, and the neighbor's is all green, you know what's going on."[4]

As spigots dried up, so did jobs. As jobs dried up, crime rose along with xenophobia; some of Botswana's affluent citizens and expatriates woke up to discover their dry hot nation had stopped being a desirable place in which to live. Looking out at the dust of what used to be lush backyards, many departed, seeking literally greener pastures, taking their firms, skills, bank accounts, and garden parties with them, just as international donors and development agencies also pulled out of the "economically graduated" country. Facing a withering and contracting economy, the government grasped at straws. Water Minister Charles Tibone toured Gaborone Dam and pronounced that the waters it held would only sustain the city for five more months, whereupon "it would have to look elsewhere."[5] Such as? If the situation continued to worsen, thirsty satellite cities like Serowe and Mahalapye would be cut off, in order to preserve the nation's capital.

This was the equivalent of Washington, D.C., announcing it would shut off water to all Maryland and Virginia suburbs. As water ran out, urban triage began, sacrificing the marginal many to save the privileged and well-connected few.

As a last resort, some officials explored the possibility of rainmaking. But there were no clouds to seed, perhaps driven off by Gaborone's urban structure. The shiny office buildings, air conditioning, hot automobile engines, kitchens, and asphalt roads all accumulated and trapped heat in the city, and scattered any moisture.[6] This phenomenon—the urban heat island effect—helped raise average temperatures from Phoenix to Las Vegas by five degrees since the 1960s, and models suggested they would rise another fifteen to twenty degrees over the next generation, reaching the point where blood vessels boil.[7]

After four decades of dutifully following foreign advice to the letter, Botswana suffered a midlife crisis. It had copied the best development practices from the United States and Sweden to the UN and World Bank. It had progressed from abject poverty to stable middle-income status. And yet it was now teetering on the verge of a dry nervous breakdown. To figure out quickly what it was still doing wrong, the country hosted an international water symposium attended by ninety experts from five continents whose countries would endure what Botswana faced right now: rocketing populations, higher demand, shrinking supply, stresses from desertification, and global warming.[8] To Botswana's horror, the upshot was that Gaborone Dam had been a grave and potentially fatal mistake: Centralized large-scale reservoirs made arid countries increasingly vulnerable to climate change.

Big, expensive water supply projects concentrated populations and increased dependence at a time when cities should disperse and decentralize demands. Water storage dams only made sense in rainy countries. In arid regions like Botswana—or, for that matter, the western United

States and southern Europe—with temperatures rising a degree every few decades, not only did dams fail to help nations endure dry spells; they actually increased the risk of drought.[9] New dams could be built, and even hooked to aquifers for recharge. But that might be pointless. By then, scientists said, rivers would have completely shriveled up, with no runoff to store. Far better, said experts, were small-scale, noncentralized, locally managed autonomous water reservoirs, where rain could be efficiently trapped, stored, carefully transported, and used.[10]

In short, somewhere out there in the Kalahari, Qoroxloo's band got it exactly right.

In crises, people look to government. Botswana's government looked up at the clouds or covetously at neighboring rivers. Mogae's top advisers expressed no public regret for terminating delivery into the reserve, but some of his lower ranking officials privately confided to me their anger at having squandered so much water from all citizens, gone forever. By September 2005, as Gaborone Dam hit a historic low at 17 percent, or nine gallons per citizen, the country grew increasingly tense. In response, Mogae asked his people to join him in praying for clouds to release their rain, known as *pula*, and for the strength to make sacrifices: use less water, treat water as a precious necessity, cultivate arid-adapted indigenous plants, recycle gray water, convert to low- or no-flush toilets, punish water waste, prevent evaporation, harvest rain, and recharge aquifers as a matter of principle. At the start of the siege, Mogae's top aides cut off water into the Kalahari reserve because, they said, "the Bushmen want to become like everyone else." Now it seemed Botswana wanted everyone else to become like the Bushmen.

As water kept vanishing faster than the government could supply it, scarcity began forcing everyone to compete, as rivals, for what little was left. Neighbors grew suspicious and hostile; nations warily eyed their shared river borders; the line in the sand sealing off the besieged Kalahari was only the latest and most prominent fault line cracking open in the arid region. Before their intense rivalry could escalate into violent confrontation, Bushmen were left two peaceful routes to negotiate their security and survival: buy water as a market commodity, or secure it as a human right.

PART V

Negotiation

CHAPTER 16

Haggling over the Source of All Life

T. Boone Pickens had an easy answer for situations like the Bushmen's. The eighty-year-old Texas oilman and corporate raider believed the fossil fuels era was over, replaced by a scarcity much, much bigger and far less replaceable. By 2025, two thirds of the world's population will face water shortages, and in expectation of that coming scarcity rivals would be inclined to drive up demand and send prices soaring. As that happened, Pickens aimed to cash in. So for eight years he quietly invested $100 million to buy up rights to billions of gallons of water—more than any individual on Earth—and then announced his plans to auction it off to the highest bidder under the timeless law of supply and demand. "There are people" he explained, "who will buy the water when they need it. And the people who have the water want to sell it. That's the blood, guts, and feathers of the thing."[1]

At the dawn of this century, few ideas generated more fierce global controversy than packaging water for private profit. To many on the right of the political spectrum, the idea of trading water as a commodity seemed obvious;[2] to those on the left, odious.[3] The Berlin Wall had barely been shattered before a new ideological barrier had arisen in its place, dividing conservatives against liberals over how to equitably and efficiently distribute Earth's real liquid assets.

As in the cold war, this emerging dry war was fought, by proxy, through ugly skirmishes taking place in hundreds of countries and in cities as different as Cochabamba, Bolivia, and Stockton, California. Qoroxloo's band was no exception. Indeed, her Kalahari microcosm revealed the fault lines of a fierce debate over the philosophical, political, economic, and practical implications of negotiating life's most precious commodity.

The core question is whether human nature is inclined to share resources like water equitably in an egalitarian society, or rather to define water according to individual "ownership" and trade it for personal gain. While Bushmen called themselves Red People based on skin color, some outsiders chose to identify them as the ur-Reds, or proto-Marxists. The Maoist Internationalist Movement, for example, cited Bushmen like Qoroxloo as walking proof that for a million years humans practiced "primitive communism."[4]

This wasn't entirely wishful thinking on the part of self-deluded social revolutionaries. Various scholars have observed how Qoroxloo remained uncorrupted by currency,[5] or how her band's system of production appeared to efficiently distribute goods and services from each according to her abilities to each according to his needs. What's more, Qoroxloo held no written property title, instead fusing her labor and capital through daily activities. Even though she never quoted from the Communist Manifesto or clutched the Little Red Book, her nonliteracy seemed only to reinforce how the weight of history was on the side of the collectivist left, who blamed all water shortages on our amoral, brutal, capitalist system. Protection of traditional beliefs by indigenous people motivated liberals who opposed "the commodification and privatization of water" and maintained "water must be free."[6]

A closer look reveals important wrinkles and nuances in Qoroxloo's resource outlook. Purely altruistic sharing did take place among the members of her extended family, a vast and intricate network of relations by blood or marriage that was often difficult for the outsider to detect or delineate. This kinship network expanded still further through a system of relationships based on the same name.[7] Yet beyond that, competition to possess important natural assets, like water, took place not just externally, between Bushmen and Botswana; intense resource rivalry was very much pervasive within Kalahari bands as well, especially in times of drought and scarcity. To reduce frequent or recurring cases of potentially violent conflict, the bands relieved tension through what could be described as informal market transactions.

Qoroxloo and her band certainly recognized customary, personal, or clan property rights or ownership. While no one codified land rights on paper, everyone kept track of tenure or who was in charge of what. People knew which families belonged to which Kalahari hunting or gathering grounds for water-rich game or tsama melons, respectively, and these relations became part of an informal, reciprocal trade matrix. Marked spears or arrows traced a kill back to a hunter who could then determine how to divide and trade his prize for other goods or services.[8] Not surprisingly, among the most precious defined and negotiated property rights were those that hinged on access to water.[9]

Kalahari sip-wells had owners, based on those who first discovered and used them. Likewise, tenure of the pan at Metsiamenong, "Vulture Water," had an unwritten title passed down through the ages to Qoroxloo. Whenever even a tiny puddle gathered in the small basin, she allowed families to rush in and grab it before any drops could evaporate, and these quantities then became the property of each household, as long as it was used primarily for drinking and cooking (bathing and washing were extravagances) and not wasted in ways that would diminish that precious liquid asset. Even plastic canteens or ostrich eggshell containers were marked by their owner so that others knew to whom that water belonged. In doing so they incurred an unspoken debt to Qoroxloo, an informal bond, but one that held the group together.

Furthermore, internal trades flourished for goods and services—including food, meat, tools, skins, weapons, tobacco, and beads—in an exchange network that bound families to bands, and bands to outside groups.[10] Qoroxloo kept exchanges in circulation and delayed reciprocity: "a return gift made too soon seemed a trade, not a gift made from the heart."[11] But while debts incurred were invisible—like holiday gifts or dinner party obligations—people kept score. Qoroxloo had a set of partners with whom she exchanged, and each partner had a different set, and so on. The Utah anthropologist Polly Wiessner revealed this extensive, semiformal and protocapitalist practice among Bushmen elsewhere in the Kalahari as *xaro*. "Xaro was one of the most powerful bonds within the social fabric, because a partnership would last for life. The process would continue. Two people would think of each other with affection, and would prepare return gifts with pleasure, anticipating the good feelings."[12]

Finally, to appreciate the possible geographic and historical extent of this complex market, consider the dispersal of Qoroxloo's ostrich eggshell beadwork. The Bushmen bead economy was an ancient form of currency, perhaps the oldest on Earth, and could be traced back millennia. At Tsodilo Hills, northwest in the Kalahari, archaeologists excavated a record of habitation stretching back from seventy years ago to 100,000 B.C. From the surface down through two meters depth—from the present to the Later Stone Age—Larry Robbins and Alec Campbell peeled away the layers of time. Where "most human artifacts and bones grew scarcer," they wrote, "broken ostrich eggshells grew abundant." One drilled bead at a deep level was radiocarbon dated back twenty-seven thousand years, suggesting a kind of crude but continuous mint.[13] Thus in Botswana were broken water canteens converted into a near universal African currency exchange, arguably the world's oldest form of abstract wealth, known locally as *!xamdzi*.

The Bushmen propensity to truck, barter, and exchange property was self-regulated by defining the terms of individual and group territorial rights: who, how, when, and where people could, for example, drink water.[14] These social systems of mutual rights and responsibilities involved property rights, entailed reciprocity, extended credit, and assumed debt. Not only did informal, subtle, and highly complex markets exist for the vital exchange of water resource goods, services, or information; they essentially formed the basis for risk reduction and a coping mechanism for survival.[15]

Her band's seemingly casual—yet in practice intricate and extensive—trade and distribution network had internalized and accounted for the exchange value of water, whether delivered by government or not. It was present whenever Bushmen managed to gather or smuggle water supplies into the Kalahari. Bushmen never expected something in exchange for nothing, and Qoroxloo regretted the day, years earlier, when her band first accepted water, from the government, for free.

WITHIN THE KALAHARI, Bushmen were prepared and willing to negotiate for water. At issue was whether they would be allowed to trade that water contractually with Botswana, at what price, and under whose terms and conditions. In theory the other side of the negotiating table should have been receptive, for more than any nation in Africa, Botswana embraced robust capitalism. Its tourism, cattle, and diamond industries all depended on global exchange, and President Mogae wanted to diversify

The tree where man was born: The drought-adapted baobab has survived millennia of harsh aridity, but now even baobabs are faltering under unnatural stress. (JAMES WORKMAN)

Survival of the driest: To adapt to radical climate change, states should honor Bushmen's ancient and resilient code of conduct, before it too is lost to the wind. (JAMES WORKMAN)

Face of Eve: Qoroxloo's DNA holds the purest genetic strains, revealing her as perhaps the closest kin to our ninety-thousand-year-old common ancestor, the earliest mother of humankind. (JAMES WORKMAN)

Water flows uphill, toward money: The Los Angeles Aqueduct pumps liquid wealth to urban California, just as Botswana reserves precious water for its three exclusive industries. (AQUAFORNIA)

Hoover Dam: Engineering marvel or vast evaporation pond? A twentieth-century hydraulic arms race pooled rivers behind forty-nine thousand large dams over the past eighty years, where waters vanish on a warmer planet. (UNITED STATES BUREAU OF RECLAMATION)

Not running on empty: America's droughts drained Falls Lake in normally lush North Carolina; regional water stress split states apart and turned neighbor against neighbor. (STEPHEN SHEPHERD)

Food is water, water is food: Qoroxloo's band obtains most nutrition and water by harvesting diverse wild foods, carefully gathered from a vast permaculture garden. (JAMES WORKMAN)

Humans coevolved with nature's order: The Kalahari captured and embedded its scarce moisture, here sealed within tsama melons, to consume or trade during the driest months. (JAMES WORKMAN)

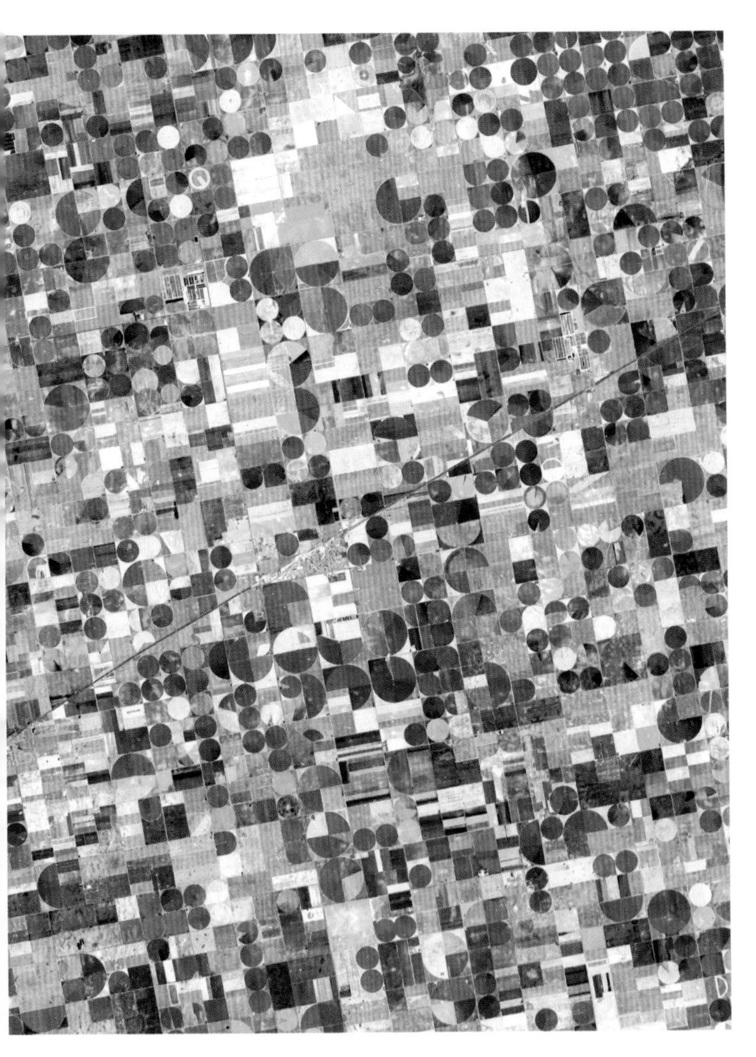

Precarious illusion: Cash crops in arid Kansas owe their life to the Ogallala Aquifer, but irrigation farmers "mine" so much fossil water, so fast, that even deep wells are drying up. (NASA)

"Water is a life force": For centuries Qoroxloo's band captured rain in a tiny pan at "Vulture Water," storing all, using little, wasting nothing, never growing too dependent on it. (JAMES WORKMAN)

Earth's oldest groundwater pump: Dispersed Kalahari sip-wells offer Bushmen a reliable form of drought insurance, but such specialized water knowledge is becoming a lost art. (JAMES WORKMAN)

Tapped out: Borehole rigs pump fossil water from below, while on the surface they break down, stir conflict, and deplete finite supplies to which people have become dangerously addicted. (JAMES WORKMAN)

Beautiful beast: Hunting is today a vanishing art, from Africa to the United States, but ancient instincts don't go away, and drought-resistant game may bring hunting back. (LIP KEE YAP)

An inconvenient meat: Botswana's bovine herds consume trillions of gallons of water; globally, domestic livestock emit more greenhouse gases than all of the planet's cars combined. (JAMES WORKMAN)

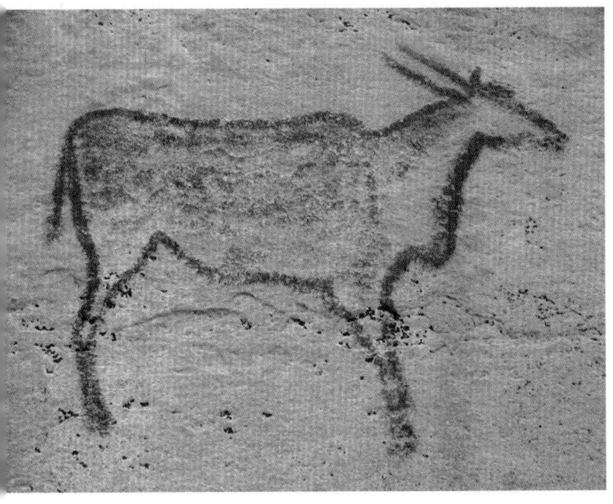

Haunted by our quest for meat: The *n!ow* in the wild eland, a water-independent prey species, drove Bushmen to smear its image on walls and to link the hunting of eland with rainfall. (JAMES WORKMAN)

Botswanan president Festus Mogae: No “Stone Age creature” allowed. “If the Bushmen want to survive, they must change, otherwise, like the dodo they will perish.” (CHRISTOPHER ALLEN BENTLEY)

Sidney Pilane: He ordered police to gas, shoot at, arrest, and jail rioting Bushmen “terrorists,” but his officials confiscated not a single weapon, except for water. (FIONA WATSON/SURVIVAL)

The Bushmen aren’t forever: As protests mounted, De Beers’ operations and brand name were also besieged by the Kalahari struggle, as Botswana tested the economic paradox of abundant diamonds vs. scarce water. (TIM MITCHELL/SURVIVAL)

Freedom means nothing left to lose: Defiant to the end, Nyare Bapalo, right, sharpens a spear with his grandson Moagi in Metsiamenong. "We are part of the game," he told me. (JAMES WORKMAN)

Living in two worlds: Roy Sesana, here trance dancing, led the fight against Botswana's government for Bushmen to remain in the Central Kalahari Game Reserve. (FIONA WATSON/SURVIVAL)

Besieged, broken, but still unbowed: Despite seven years of torment and water deprivation, the last men, women, and children of Qoroxloo's band at Metsiamenong have never caved in to government pressure. (JAMES WORKMAN)

Setheke on Qoroxloo: "She raised her kids in the dry times when there was no water whatsoever, but she managed to keep them and raise them. All that because of her skills." (JAMES WORKMAN)

Homicide or suicide? Mongwegi shows how he found Qoroxloo killed by shock and dehydration after four days alone, isolated by armed officials, concealing the dryness within her heart. (JAMES WORKMAN)

Images of power: Mystery surrounds rock art like this handprint at Tsodilo Hills, a place in the Kalahari with hidden waters. A haunting smear, it seems to proclaim: "I endured." (JAMES WORKMAN)

his economy further by removing all shackles to commerce and competition. The one exception to his market fundamentalism, it seemed, was water.

When it came to water, Botswana practiced socialism. It endeavored to supply freshwater to all citizens, regardless of where they lived or how empty their pockets. Since 1978, neither sand nor heat nor bitter drought kept the government from delivering water to its poor, for free. With justifiable pride, the country boasted 100 percent access to safe water in urban areas and 91 percent access in rural areas.[16] But as socialized water grew increasingly expensive, hauling water into the Kalahari began to drain the government budget. This became a prominent rationale in the legal battle between Bushmen and the government. "It was costing us thousands and thousands of pula (Botswana's currency) every month," bemoaned Mogae's advocate in court, Sidney Pilane, emphasizing the country's reputation for fiscal budgetary responsibility. "It was unsustainable providing [water] services. Have you been to the Kalahari reserve? Because if you've been there, you will know that it has to be one of the most wicked terrains anywhere in the world."

Wicked it was. A borehole breakdown took weeks, and thousands of dollars, to repair. Hauling water over an unforgiving landscape with specialized fuel, drivers, and equipment all added up. The government ticked off rising labor, fuel, pumping, loading costs; costs of transport and of repair; costs of overtime, benefits, and hardship pay. There were also opportunity costs to consider, since time and money spent taking water to Bushmen meant time, money, and water taken away from someone else, someplace else just as thirsty and needy, so much so that officials claimed it retarded developments in other parts of the district and that it was cheaper for the government to pool its resources in one village. The growing costs against a shrinking operations budget meant socialized water deliveries into rural areas were no longer feasible.

Actually, it is hard to find where it might still be feasible. The expense of water delivery—filtering, piping, drainage, sewerage, and treatment—has exploded everywhere on Earth, bankrupting governments, including America's. Cities in New Jersey, Indiana, Wisconsin, Louisiana, Georgia, Texas, and California have grown so saddled with so much debt from subsidized water that they can't maintain or repair (let alone improve) existing pipes and plants to meet escalating demand. Ballooning public water expenses have outstripped U.S. current spending by $500–$800 billion.[17] Because of this, thirty-six states now expect to suffer water shortages in the next decade. Worse, a third of all U.S. water utilities have ancient pipes, a creaky infrastructure, and no funds to fix anything.[18] Within and between regions, delivery failure is literally pulling apart the United States at its leaky, watery seams.

As a frugal executive, President Mogae took pride in budgetary rigor, fiscal austerity, and monetary restraint. Ever the vigilant economist against bloat and waste, he decided that water deliveries into the Kalahari, and there alone, had become too wasteful. Mogae cut off the Bushmen because he could not afford the high price of their free water.

So be it. If socialized water delivery rose too high for the government, the Bushmen determined they would somehow try to cover costs of pumping and transporting their water by themselves, and prepared to negotiate. They found charitable public and private backers willing to cover the full cost of water delivery to let Bushmen stay in the reserve. With secure funds, their offer was brokered through a formal proposal by the European Union. Botswana could be absolved of fiscal liabilities, invest its development funds elsewhere, and wash its hands of water-related costs. All parties weighed the prospects of securing water peacefully through a free market transaction, just as T. Boone Pickens envisioned.

But it turned out to be less simple than he, or anyone else, could have imagined. A thorny issue involved the correct price of water. To determine a fair opening bid, Qoroxloo's band and its broker asked by how much they had drained Botswana's federal reserve. Over the next five years, despite countless requests in court, the government never did answer.[19] Under socialized rural water, few really cared, and no one kept records. Botswana was not alone in its ignorance about where to start. At that very moment eighty water-stressed countries were engaged in similar negotiations over a $1-trillion market, all haggling over the transaction price of water.

The European Union was Botswana's biggest export market, trade partner, and aid donor, so its offer could not be ignored. If President Mogae accepted the deal, he would gain considerable time, labor, money, prestige, diplomatic relations, and international support. As an International Monetary Fund veteran, Mogae was predisposed to the beauty and efficiency of markets and enforceable contracts. Like Pickens, his former IMF colleagues pushed the one solution that would improve delivery, eliminate scarcity, and allocate water fairly and responsively and durably in a way that was economically painless and politically invisible.[20] To systematically improve Botswana's waterworks, Mogae merely had to break up the state's oldest and biggest absolute monopoly and embark on a water delivery privatization contract.

Mogae had already privatized the country's only seasonal river for De Beers; privatized boreholes for cattle posts; privatized water for tourist ventures; and contracted with a private corporation in a north-south water transfer scheme. So he was not inherently afraid of water contracts. But these precedents all involved trusted partners in symbiotic ventures that Botswana could control. Unpredictable and disloyal Bushmen, by contrast, could emerge from this deal with more autonomy. If Botswana privatized water deliveries to the Kalahari, it might come under pressure to do so a second, third or fourth time to other minority interests, fragmenting the country, diluting the president's authority.

Ever cautious, Mogae deliberated every issue, especially one so radical as a contract that put a price tag on water. Before making a decision, he sought out the actual experience of other countries, especially the United States.

IN 1758 BENJAMIN Franklin wrote, "When the well's dry, they know the worth of water." But in truth he was as baffled as Adam Smith by water valuation. Franklin's own wells never ran dry, and his entrepreneurial mind steered clear of buying, selling, or trading the wet stuff. He left water privatization to two more ruthless Founding Fathers.

By 1799, New York City was booming on a parched and pestilential island surrounded by water too brackish to drink; its population knew the worth of water, all right. Imports of fresh "tea water" cost more than beer. As demand grew, an opportunity presented itself. That year Treasury Secretary Alexander Hamilton and Vice President Aaron Burr—before they became famous rivals and infamous duelists—tried to corner a lucrative market as partners in an unprecedented private venture known as the Manhattan Water Company. Burr and Hamilton negotiated a government contract in which they would provide water indefinitely to a thirsty isolated population. They promised to tap into a pure water source of the then-pristine Bronx River from upper Westchester County and divert it via aqueduct to Manhattan, thus quenching the thirst of the thriving metropolis. Their venture would, of course, require exclusivity as a "natural monopoly"; they couldn't very well have competing pipes from the same source constantly ripping up streets and crisscrossing the city. Oh, and they'd need a contract in perpetuity to justify up-front capital expenditures. Finally, they would charge a modest fee, enabling them to invest profits as the executives saw fit. On April 2, 1799, the influential titans won their charter and set up company offices at what is now 40 Wall Street.

Almost immediately, private water ran its own course. Burr and Hamilton's new company cut costs at every opportunity. Instead of tapping the distant Bronx River as promised, it simply dug and pumped a cheap well on the upper half of the island. Instead of its promised elaborate network of sealed pipes and vast reservoirs, it stored water in a large iron tank and constructed twenty miles of pipes from cheap, hollowed-out logs hinged together. Instead of servicing New York's 60,000 thirsty people living in the city in 1800, it sold 700,000 gallons of water to the 1,400 most affluent, privileged, and literally well-connected households who could afford whatever rates the company felt free to charge. Having banished competition, the company reaped considerable profits, none of which were reinvested to improve water quality or services. Instead, Burr leveraged the company equity, capital, and income into making secured loans with which he established an entirely new bank—later Chase Manhattan—to rival Hamilton's Bank of New York.

The Manhattan Water Company answered to no one, but rather hovered above the young Constitution as an entity unto itself, mutating into a corrupt and incompetent monopoly whose ironclad contract could not be broken until cholera and fire drove enraged citizens to take back public control of their water supply three decades later.[21] The ill-fated venture drove swing voters from Jefferson's party and may have even played a small role in the 1804 duel that snuffed out the extraordinary man on today's ten-dollar bill.

Such behavior did not bolster the case for contracting out private water for Bushmen, or anyone else. Still, that was two hundred years ago; surely the leaders of the free world would by now have worked out the kinks of water privatization. As it happened, just as Mogae was weighing the EU offer, President George W. Bush's "ownership society" sought privatization of America's waterworks while his Congress wrote the pro-privatization Water Investment Act. Senators updated obscure provisions in the 1994 NAFTA[22] so that all three countries—the United States, Canada, and Mexico—would convert water into a private commodity subject to the rules and discipline of the unfettered market. Fortune 500 companies worked with the World Bank to lobby governments to transfer some or all responsibility over the assets and operations of public water systems into private hands.[23]

Wall Street salivated, for obvious reasons. Imagine the venture capital pitch: dwindling supply; soaring demand; open entry; no competition; and total price inelasticity due to a captive customer base growing into a possible market of 9 billion clients. Few could resist. Entrepreneurs schemed up everything from dragging icebergs to capturing fog, as brokers scrambled for a slice of the action. Water securities and stocks expanded in value, number, and complexity, as T. Boone Pickens and the Bass Brothers competed with online brokers like WaterBank.com, iAqua .com, and WaterRightsMarket.com. In five years, global bottled water revenues doubled to $15 billion. Investors grew giddy. Unfortunately, most of the best private opportunities already had been consolidated into the hands of a few water service industry conglomerates: Germany's RWE; France's Suez and Vivendi/Vitel; America's Enron and Bechtel. Collectively these companies alone brought in $200 billion in revenues and grew at a rate of 10 percent a year.[24] The smallest of these institutions eclipsed Botswana's financial reserves and employed more people than Botswana's entire government workforce. Each served between forty to ninety times more customers than Botswana had citizens, and they were just getting warmed up. Private water management was projected to hit $1 trillion by 2015. Analysts predicted that multinational control of local water systems would absorb 1 percent of the world's population each year. Market watchers enthused about "an explosive growth business that presently offers investors a shot at a couple of double-your-money stock ideas" because industry trackers "caught early sight of a major global risk: a burgeoning worldwide water crisis that is now a reality."[25] Pension funds embraced water stocks because they involved secure, thirty- to fifty-year contracts. And portfolio managers viewed water as "socially conscious investing, an opportunity for people to help the world solve a major and critical problem."[26] Corporate boosters welcomed water scarcity as "one of the world's great business opportunities," one that, better still, "promises to be to the 21st century what oil was to the 20th."[27]

Or not. Manhattan's first disastrous experiment two centuries ago set a pattern followed later by other cities, first in the United States, then abroad. In Atlanta, United Water won a contract promising to make the city an "international showcase" for public-private partnerships. Instead, the Suez subsidiary failed to repair major water-main leaks or fire hydrants on schedule, violated federal drinking-water standards, neglected to collect on some bills while cutting off water services at apartment complexes that had fallen behind in payments.[28] Hundreds of residents complained of brown water with particle-laden debris floating in it; new mothers learned of a "boil only" alert only after feeding their babies milk formula.[29] The following January, after a six-month audit, Atlanta unilaterally ended the contract sixteen years early, retook control of its water system, and kicked the company out of town. The city's motto had been "Atlanta grows where water goes," and the commissioners decided that they would "prefer to see the city in charge of that destiny."[30]

Private water monopolies held obvious and inherent dangers: no incentives to deliver to the poor; no competition to improve and maintain water quality; no costs for extracting from a public good; no reason to invest profits back in the local resource; no cause to address public risks from health, sanitation, or fire. Time after time, private water monopolies led to chaos, riots, distrust, and pandemics.

Most such ventures have failed for one of three reasons. The first was cost. Unlike public utilities, water companies had to clear a profit margin and they had to pay taxes. Together these added 7 to 100 percent higher operation cost, passed on to customers. To maximize returns, companies gravitated toward denser, richer populations while avoiding laying pipes in urban slums and rural backwaters. When the poor couldn't pay or got left behind, they sent a nonmarket signal: they cast their vote against privatization.[31]

Next was control. Few citizens feel comfortable with foreign hands on local taps. If a government like Botswana can't run an efficient public water system, it would be even less likely to effectively regulate a private conglomerate twice its size, based in another hemisphere, ensconced behind three corporate shells. For example, one local Monterey, California, enterprise was part of the state company Cal-Am, which was held by the national American Water, which, in turn, was owned by publicly traded multinational RWE in Germany, whose distant CEO was more directly responsible to Singapore shareholders than to Carmel customers. And while public ratepayers circulated money locally; private customers sent profits overseas.

Finally, there was performance. If only one monolithic entity could control the entire conveyance system for twenty to thirty years, it mattered little if the entity was public or private: Both excluded rivals. "Water privatization is not about competition," concluded Peter Gleick in an independent study about the new economy of water. "These are long-term monopoly contracts. This isn't free enterprise or a competitive market."[32]

Two centuries after the Manhattan Water Company was established, the same types of riots and diseases and failures once again translated into what private equity analysts described sadly as a "chilling effect" on the water industry.[33] Voices for flat-out privatization of water grew subdued. American cities and developing nations began to reject or cancel proposals to shift public water responsibilities over to business. If anything, the growing antiprivatization forces were increasingly on the counterattack, demanding that private control of water must be restored to federal governments acting on behalf of the rural poor and marginalized indigenous people.

But therein lay the supreme irony in bargaining over the price of water. Liberals were demanding that governments like Botswana must protect people like Qoroxloo from privatization contracts, but at that precise moment Qoroxloo's band sought a privatization contract to protect themselves from Botswana's government. By taking water out of officials' hands under the EU contractual offer, Bushmen were seeking no profit, no monopoly, only better quality, health improvements, and total accountability. As a microcosm of the world's neglected people, Bushmen knew the real worth of water.

The voluntary and reciprocity-based water transactions long practiced in the Kalahari were a far cry from the monopoly privatization contracts enjoyed by Suez or Bechtel. Indeed, it was only in those unprofitable corners of the world bypassed by formal public-private contracts—the rural backwaters and urban slums—that genuine supply-and-demand forces could emerge. Among various competitive water vendors, the poor negotiated the price of water every day. Since domestic water responsibility fell mostly to women, females paid the predominantly male vendors with money, time, energy, bartered goods, or sexual favors. This was a free and unregulated market of the oldest kind. Such an informal, decentralized water resource exchange had been going on since the dawn of time, forging an autonomous kind of self-governance, a hydro-democracy unto itself, bypassing the need for a top-down authority, and thus posing an implicit threat.

Most governments tolerate these informal water exchanges, known as gray markets. Botswana outlawed it as a black market. Usually, officials are eager to tax economic exchanges in order to generate revenue for governance. But to enter the formal economy, water contracts and exchanges would require title ownership of water, and that, in turn, would shift power and authority down and away from the government. So while the EU contract would have improved water deliveries in the Kalahari at zero cost to his government, Mogae ultimately rejected the offer.

When the missed opportunity for a water contract later arose in court, the president's adviser, Sidney Pilane, sneered: "We do not need Europeans telling us what to do and what not to do. They can give them all the money they wish but we really don't care, and we resent their involvement in our affairs. We are concerned with our own people, we will do what we think is best for them."

It seems more was at stake than a transparent and mutually beneficial exchange. To the government, cost didn't matter. What did matter was state sovereignty and resource ownership and who could access water in the first place. In a free world as wells ran dry, thirsty humans could and did haggle out the worth of water among equals. But Bushmen were not considered equals, and so long as they couldn't legally own Kalahari water, they had nothing to trade and so could not bargain.

President Mogae chose that outcome. Evidently his rejection of a peaceful private market transaction had nothing to do with water's price and everything to do with water's control.

CHAPTER 17

Human Rights, Water Wrongs

Now prohibited from negotiating contracts to secure her water as a private commodity, Qoroxloo's band was left no choice but to seek it aggressively as a fundamental right.

This was legal terra incognita, and human rights lawyers initially filed the lawsuit in early 2002, hoping to reverse the evictions, gain leverage, bring all parties to the table, and broker a fair settlement. "The government should not feel boxed into a corner," one local attorney told me on several occasions. But when the president's officials established their siege of the reserve and refused to budge, 243 Bushmen challenged President Mogae head-on in Botswana's high court.

Many expected a swift judgment, but instead the case crawled across 251 weeks like a Kalahari tortoise at midday. Stenographers churned out 19,000 pages of court transcripts. Bushmen plaintiffs and government respondents filed 4,500 pages of legal documents. The legal process was agonizing, and the trial only got under way in 2004, whereupon the first Bushman witness, hunched in the witness stand, spoke softly. Too softly. His voice was nearly inaudible. Within minutes an irritated Chief Justice Maruping Dibotelo had him stop mumbling. "You must speak up!"

Amogelang Segootsane explained his voice was naturally low.

Dibotelo leaned forward, instructed the witness to stand on his feet and project from the abdomen so that everyone could hear.

Amogelang said he was exhausted, having traveled a long and difficult journey on foot through the desert to get here. The city was disorienting. He had camped out in unfamiliar bush and had not slept well.

Dibotelo repeated his instructions for the third time.

An awkward silence followed. At the country's defining human rights trial, the court was demanding that a thirsty, destitute, fatigued, and frightened witness stand up for several hours in a hot and airless room under cross-examination by a sneering government attorney while officials poured ice water from pitchers in front of the man who for two years had been denied a drop.

Dibotelo paused to consider the situation.

The United States was already accusing Botswana of gross human rights violations against Bushmen: violence during interrogations; lengthy judicial delays; limits to journalists and academics; harassment of activists.[1] Of course America itself faced similar allegations: excessive force during questioning of suspects; holding prisoners indefinitely without trial; press restrictions; and using water to extract information.[2] But if human rights attorneys challenged the Bush administration for waterboarding,[3] Mogae's government was on trial for precisely the opposite reason.

Segootsane said he found himself standing up, "here in this box," ordered to project his voice, for two reasons. First, the government had cut off his family's regular supply of drinking water. Then it had stopped him from bringing a regular supply of drinking water to his family by himself. He didn't want to come to court, but he had no choice.

Dibotelo stressed that this was not an inquisition. "We are not trying to persecute . . . torment you . . . you can sit down and rest when you feel the need."

Witnesses who were subjected to waterboarding typically gave in within fourteen seconds,[4] but water deprivation took longer. Some Bushmen endured months or years of thirst before caving. A few dozen Bushmen lived on indefinitely or died under questionable circumstances. But they never cracked. Still, state-sponsored thirst eventually accomplished the task at hand and offered undeniable advantages to those in control: no scars, no witnesses, no direct force, no physical restraints, and no apparent liability.

The high court had to decide whether that coercive method—which might be called the intentional use of compulsory thirst—was legal. Judge Dibotelo offered the Bushmen plaintiff a glass of water as a courteous gesture. But could his government deliberately restrict or prevent Bushmen from access to water? The question was not hypothetical.

Repercussions from the high court's precedent-setting ruling would resonate far beyond Botswana's borders. On behalf of 6 billion humans, the UN danced around the very same question: Were Qoroxloo and all other Bushmen inside the Kalahari reserve endowed with a human right to water?

For that matter, was anyone?

Liberals generally held that truth to be self-evident.[5] At the dawn of this century, a loose assembly of antiglobalization protesters, trade unions, religious leaders, public utilities, peasant farmers, American social activists, French intellectuals, and human rights groups had galvanized into the self-proclaimed global water movement.[6] As *the* essential element without which no living thing can exist, the group's leaders, like Maude Barlow, argued that water must be secured for the people, by the government, against big business.[7] And its manifesto demanded: "The Earth's fresh water belongs to the Earth and all species, and therefore must not be treated as a private commodity to be bought, sold, and traded for profit . . . the global fresh water supply is a shared legacy, a public trust, and a fundamental human right."[8] Armed with right against might, the global water movement provoked nonviolent confrontations and proceeded to chase "foreign economic imperialists" and "water barons" like Coke, Vivendi, Suez, and Bechtel out of town, from Kerala, India, to Buenos Aires, Argentina, to Sydney, Australia, to, most spectacularly, Stockton, California, where citizens rose up to overthrow a $600-million water privatization contract with the foreign-based OMI-Thames.[9] Eventually RWE, the German parent conglomerate of a dozen water company subsidiaries from coast to coast, fled the U.S. market altogether. Finally, the global water movement called on the United States, World Bank, World Trade Organization, and United Nations to insert key phrases their founding charters had left out: equal public access to rivers, lakes, and aquifers; equal shares of public water to drink, wash, and bathe; and the inalienable right to water.[10] "Basically we see water as an issue of human rights versus corporate rights," said Barlow. Indeed, she added, "water is the most important human-rights issue of them all."[11]

Conservatives adamantly disagreed. This so-called right didn't hold water, figuratively or literally. America's Founding Fathers were not socialists. They would no more have engraved in the Constitution a right to water than they would a right to land, food, medicine, jobs, housing, transportation, or fuel. Doing so might even weaken other human rights by making people increasingly dependent on big government. Certainly, water was a necessity. But nothing good came from calling water's economic goods and services a "right."[12] To secure access to water, people must simply employ "real" and "classic" political rights like free speech, free assembly, and free press. Indeed, "the trouble with rights like 'water and sanitation' is that they often achieve the exact opposite of their aims because they invite state intervention into all kinds of areas. Thus, these rights run the risk of bringing about exactly what human rights are supposed to prevent: an omnipresent state."[13] What's more, ran the counterargument, it's impractical. How would any emerging so-called right to water be quantified? Would people get an unlimited supply? Would it flow as unrestricted as speech or religious worship,[14] or would failure to pipe free water to every door, on demand, expose leaders to prosecution for human rights violations? What Thomas Paine said about liberty—"What we obtain too cheap we esteem too lightly"—could equally apply to water. Instead of ensuring conservation for all species, said conservatives, a human right to water would quickly lead a nation to waste, pollution, corruption, biodiversity extinctions, and, quite literally, state insolvency.

Between Left Bank and Right Bank, billions of nonideological people like Qoroxloo and Amogelang fell through the cracks. For example, Bushmen did not oppose water as a tradable good, but that conservative option had been closed off, and when denied access to water, the so-called real rights to life, liberty, and the pursuit of happiness had consequently been infringed. At the same time, Qoroxloo found no liberal written precedent, either. Paine's *The Rights of Man*, Jefferson's Declaration of Independence, and Madison's Bill of Rights were all silent on the matter; even the postwar UN Declaration of Human Rights failed to mention freedom from thirst. At the time of the siege, no country anywhere recognized, enforced, and clearly defined an explicit human right to water. And against the global water movement, one powerful country fought quietly to keep that issue off all multilateral agendas, out of written charters, and banned from binding statements. It wasn't North Korea, Burma, or Cuba that smothered debate about the human right to water; it was the United States of America.[15]

The United States attended multilateral UN meetings with the express intent to water down language that elevated water as more than an economic good. The richest, most powerful, and most individualistic country in the history of the world did not recognize water as a human right, and wanted nobody else to, either. For years, the legality of thirst remained an ideological abstraction, unprovoked and untested in court until the challenge from Bushmen starting with Qoroxloo's low-voiced coplaintiff, Amogelang Segootsane.

When the government's convoy had come, Amogelang recalled being surprised at "how much water was poured out of the tanks." He told the court he "did not know what to think," but assumed "there was something wrong with the people's heads, or the tanks." The intent soon became clear. One truck carried the tank away; others carried off his neighbors. Those who remained "were very hurt." Their provisions dwindled. As a husband and father of three, he had to act. If the government could not bring water to his family, he would.

So one day he stored up wild *kgengwe*, a water-rich plant, for his family, and proceeded to walk south. He crossed tiny salt pans. Well outside the Kalahari reserve, he filled plastic barrels with water at a tap and brought them back in a borrowed donkey cart. He did this every few months until the day he was blocked. As the guards made him pour all his water out that day, they explained they were only following orders, and if he didn't like it he could write their bosses, asking special permission. Amogelang could not write or count past ten, but he knew who could, and decided to seek her. He walked farther out of the reserve to Kaudwane, slept near a fire with people he knew, and told them he sought permission to take water into the Kalahari. When they asked if he could also bring water to their families remaining in Metsiamenong, he said he did not know, but would try.

Amogelang rode south, sharing a bareback horse until he arrived at Letlhakeng, a town so large it had a gas pump. From there he hitched a ride over smooth asphalt until reaching Gaborone, where drivers killed more people each day than Botswana's lions killed each century. He could not read signs but searched the disorienting streets. He asked directions, in his low voice, and pronounced a name. People knew it. They pointed him toward her understated office, where he stepped up to the door and knocked.

ALICE MOGWE WAS a respected, no-nonsense, progressive liberal activist. Since earning her law degree she had quietly, and more often not so quietly, made a name for herself as an attorney, eventually addressing the UN. A decade earlier she had founded Ditshwanelo,[16] or "human rights": the prism through which she saw her homeland, her people, her mission. Any fight for rights invariably embraced the downtrodden underdog. Her ideal client might be an abused rural female HIV-positive Muslim communist gay Zimbabwean refugee. Reaching beneath all these outcasts, she defended Bushmen or, in her language, Basarwa.

As a local maverick from a royal tribal family, Alice was uniquely positioned to do so. Only she, not a foreigner, could educate her countrymen about the insidious nature of torture used against Bushmen hunters. Through Ditshwanelo she could legitimately investigate and challenge their underclass status as squatters in their own country. Alice knew the language intimately enough to trace origins of *Basarwa* to a corruption of *bao ba-ba-sa-ruing dikgomo,* which is to say, those who do not rear cattle, and then scold her nation for defining Qoroxloo in the negative, and abnormal, in terms of what she lacked.[17]

Trouble was, Alice's crusading organization could barely stay afloat. As foreign funds dried up, quixotic charities like hers might have to court the favor of government and actively seek out senior political figures for help, the same figures she might later need to challenge. It was a frustrating quandary. Alice sat at her desk with a back support staring at a wall of posters filled with worthy battles she had no time or money to fight. She firmly believed the lawsuit involving Qoroxloo and 242 other Bushmen had been inexcusably delayed by the aggravating rhetoric of foreigners, and now those same overseas human rights groups—unburdened by financial constraints—had taken the case out of local hands to fight in their typically Western confrontational manner. Alice, by contrast, still believed fervently in quiet negotiation. Then again, she had no choice.

The receptionist told her she had a visitor. She heard who it was and knew how far he had come to reach her office, and she stood to welcome him. He was still dressed in her husband's hand-me-down clothing that she had provided years earlier.

"Dumela, Rra," she said, greeting him as an equal.

"Dumela, Mma," he replied, smiling back.

"What can I do for you?"

He wanted to tell her his troubles, but she knew them. He wanted to convey his hopes and fears, but she shared them. So he cleared his low, barely audible throat, hoarse from the dusty journey, and said "We have no water."

BOTSWANA HAS ALWAYS MAINTAINED IT never used force in the central Kalahari. When confronted in court with hard evidence of how, acting on orders, the president's subordinates had most definitely deployed compulsory thirst in its deliberate efforts to make Bushmen move, the government attorney Sidney Pilane vigorously denied that any official had ever deliberately ended, stopped, destroyed, cut off, or terminated Bushmen water. Those words sounded so cruel and brutal, so—terminal. What the government merely had done, he asserted, was to "move its water provision" from one place to another.

It was a farcical legal claim, and a clever one. But before it could be tested, the argument left open a loophole that lawyers like Alice could exploit. She urged Bushmen to accept water in the new place outside and then bring it back to the old one inside.

The government hadn't figured on that response. But as part of its siege, Botswana's attorneys found yet another legal rationale that would try to prevent it. No one could interfere with government policy; policy was based on denying water exchange; so officials halted all trade across the reserve's boundaries. Water, along with anything else, became legal contraband.

Whereupon Alice found a second loophole. By definition, no individual can trade goods or services by oneself. So Bushmen women and men inside could go out and haul water back to themselves.

Officials apparently hadn't considered this possibility, either, but they soon were forced to. On behalf of all Bushmen, Amogelang requested permission "for us to enter the [Kalahari reserve] with water. So that we may have something to drink everyday. The places to which the water will be taken is Kukama, Metsiamenong and Gope. It is really heartbreaking when one sees the sick orphans and the pregnant women."[18]

The next day, Botswana's Water and Wildlife departments passed the buck—"We have come to the conclusion that it is not our responsibility to give permission to people to carry drinking water"—and referred the issue to the Ministry of Local Government.

Five days later, the Ministry of Local Government's permanent secretary explained calmly how his office did not "implement regulations relating to parks." It operated under the fiction that no one remained inside the reserve; holdouts stayed of their own volition, in no-man's-land, and were "not the responsibility of local government."[19]

As an exhausted Amogelang sat before her, Alice had to explain how the Wildlife Department would let him bring water to his family once local government signed off, except local government couldn't sign off because it had no authority over Bushmen once they entered the Kalahari reserve; local government would quickly sign off on Wildlife and let him carry water inside, if he and Bushmen inside the Kalahari reserve left; only in that case permission would not be necessary because they would have moved outside the reserve, to where the water was. Such circular logic infuriated Alice. She believed in Botswana, took pride in its peaceful traditions, and strived to improve its governance nationally and its reputation globally. As the water situation deteriorated and options ran out, she tried to turn crisis into opportunity and give diplomacy one last chance.

Minister of Local Government Margaret Nasha might be described by generous authors as "traditionally built" and by everyone else as fat. Nasha was the official who, along with the military officers, was responsible for cutting off water to Qoroxloo's band; to Bushmen, she was the corpulent embodiment of their rival. Bushmen said Nasha spoke down to them, as helpless children in need of guidance. Nasha compared Bushmen with elephants needing to be culled. Bushmen loathed Margaret Nasha. Alice picked up the phone.

She requested a few minutes with Nasha for a quick talk about certain unforeseen aspects of the Kalahari reserve situation, with no direct bearing on the court case. Nasha knew how Alice's tongue could get started and never stop, so had scoffed, only partly teasing: "You? Quick talk? Won't take long? Huh!"

When Alice showed up with Amogelang at her side, Nasha visibly stiffened, and her eyes narrowed, but she held her anger in check and gestured for Alice to say what she had to, face-to-face.

For a change Alice said little, instead turning to Amogelang. "Why don't you tell her what you told me?"

He looked at Nasha and again in that low, soft voice said, "We have no water."

Nasha came uncorked. According to two of the three people in the room, Nasha proceeded to excoriate Bushmen like him, who remained inside the reserve, correcting him that there was water, plenty of water, because the government had offered water, more water than anyone needed, schools with water for children and water for everyone who wanted to develop like all citizens all over the country, until, at last, she ran out of steam.

Then both women turned to Amogelang for his response, and he repeated what was at stake for billons who shared his predicament.

We have no water.

WHEN BOTSWANA CUT off Bushmen water in 2002, few had heard of a "human right to water." Three years later much of the outside world, from France to India to Ecuador and South Africa, was taking steps to make that right explicit.[20] Bowing to "a growing movement to formally adopt" it, the Vatican proclaimed, "The right to water is thus an inalienable right." Even water-intensive industries like Nestlé and Coca-Cola—which in theory would face restrictions on economic activities, a weakening of demand for their product, and a potential hit to their bottom line—called for recognition of a human right to water for the sake of certainty and preserving their brand name.[21] Finally, in a statement backed by Kofi Annan—and opposed by the United States—the UN Committee on Economic, Social and Cultural Rights called water "indispensable for leading a life in human dignity. It is a prerequisite for the realization of other human rights . . . The human right to water entitles everyone to sufficient, safe, acceptable, physically accessible and affordable water for personal and domestic uses."[22]

The UNspeak would have aided Qoroxloo and her band if the words had been legally binding, and not just legally hinting. The UN outlined what state parties like Botswana should, could, might, and really ought to do, if and when it could find the time.[23] But for all the eloquence, the UN's statement lacked teeth. There was in principle an implicit human right to water.[24] Explicitly, that right did not yet exist.

Back in Nasha's office, as citizen, advocate, and government official squared off over the one resource they each shared and all needed in order to survive, it was hard to imagine a more subversive idea. Amogelang embodied the moral imperative, Alice provided the legal context, and Nasha, who had to govern, sent them away and pondered what action to take. She put her finger to the wind and made a few calls. Days later Ditshwanelo, Alice's human rights organization, received a letter from Jan F. Broekhuis on behalf of the director of wildlife and national parks: "We are pleased to be able to grant you permission to carry water into the Central Kalahari Game Reserve for use by yourself and your immediate family . . ."

Alice was thrilled and cited this as a perfect example of how one-on-one negotiation trumped the polarizing Western hard-line confrontational approach. Now Amogelang could continue his long donkey cart trips. It seemed a victory, a vindication of quiet diplomacy that affirmed the emerging human right to water, in writing.

Or did it? The letter ominously concluded ". . . until further notice. Note that this permission does not permit you to supply water to any other persons that may reside in the Reserve." Why "until further notice?" What exactly did "immediate family" mean to someone with twenty cousins, in-laws, nephews, and nieces? And what happened if he did supply water to Qoroxloo's band, in Metsiamenong, as he had promised? The phrase "grant you permission" was a far cry from "recognize your inherent God-given right."

Alice protested that Bushmen should not be required to beg permission to bring water wherever they wanted from officials who were engaging in unlawful conduct. But Nasha's government decreed otherwise. In demanding the last word, it used variations of *permit* and *permission* five times in three sentences. That sent a signal, loud and clear. Like a driver's permit, water extended as a temporary license need not be defended as an unconditional right. Accordingly, Botswana's government could choose to grant what people desired, but it was not obliged to protect a right with which people were endowed.

The difference was subtle but profound, for the scales could always tip back. At any moment, the privilege that the government bestowed as a courtesy could be temporarily rescinded or permanently repealed. Something given could be taken away.

Two years later Botswana proceeded to do just that. The government alleged, without evidence, that Amogelang had been hired to bring water into the Kalahari, thus breaking the terms of its generosity. Officials reasserted that Bushmen could either stay inside without water or move outside to get water, but could not traffic back and forth carrying water of their own. A final letter concluded, "the aforementioned permit has been suspended until further notice," and denied Bushmen freedom to fill up tanks and return home. Permission for water was revoked.[25]

Amogelang's extended family was subsequently forced, for the first time in their lives, to depart their ancestral homeland. From the day of the cutoffs they had lasted three years, two months, and eight days before finally caving in to compulsory thirst and state-sponsored dehydration.

Alice continued to negotiate legal terms with the government, but kept hitting her head against arbitrary rules of state officials, who claimed to be acting on the larger interest of Botswana. At one level, the UN became even more assertive in its statements about water, but failed to walk the walk. It remained for Botswana's high court to rule whether Bushmen deserved access to water as an unequivocal human right, on their own terms, in their own land. Yet even that court's rulings could be nullified by those with power.

Perhaps human rights merely reflect the congealed politics that result when various parties of equal raw power grind down to an uneasy peace. Indeed, some lawyers and scholars trace the birth of human rights to a similarly temporary truce brokered eight centuries ago during a dispute caused in part by who enjoyed access to water. In the thirteenth century, Britain's King John fenced off streams, blocked river navigation, and sold monopolies to water resources that used to be free for all. He restricted water access until subjects revolted in a medieval asymmetric war. Thirsty serfs put pressure on their feudal lords and barons, who in turn made the king restore access to water for all, until "the rivers that [he] fenced were directed to be laid open."[26] They forced his hand at Runnymede—notably, an island within a river, and thus owned by no individual—to sign the Magna Carta.

Thus scarcity brought conflict until a powerful equilibrium led all sides to inscribe the foundation of human rights. These came not from God, not through reason and conscience, not jotted down by NGOs to be passed by UN resolution, and not, as Americans were taught, conceived in liberty and born immaculate.[27] All rights—and limitations on the state—emerged through ugly and messy processes, repeatedly clawed and scraped and forced into the light, where they would have to be defended in perpetuity.

Until that happened, tensions continued to escalate. Qoroxloo's stubborn band was the last of those who never caved in to the government's compulsory thirst, who never surrendered to the siege, and who, as a consequence, brought armed officials to advance on their camps. As Mogae constricted his line in the sand, death would come even to Bushmen denied access to water as a human right.

PART VI

Confrontation

CHAPTER 18

Primal Instincts and the Realpolitik of Water

DROUGHT CAN EXPOSE THE DARK SIDE within mammals of the African savanna and bring latent aggression to the surface. Shriveling water holes cause abnormal behavior even among supposedly exclusive herbivores. During droughts, I've seen hippos gnaw carcasses and thirsty zebra kick to death a baby wildebeest while anxious adult elephants crush life from one of their own infants. But of all Africa's water-dependent beasts, one parched animal stands out as particularly, and calculatingly, lethal.

The thirsty primate carefully plots when and how to attack. When Kenyan relief workers opened a water tanker truck for drought-struck villagers near the Sudan border, vervet monkeys gathered in the trees, descended from branches, and for the next two hours fought, were wounded, and died for access to the last precious drops.[1] Chimpanzee and bonobo—which most closely share our own DNA—selectively attack their rivals over access to vanishing resources. Yet of Earth's 233 primate species, only two left the forests to endure and inhabit the arid African savanna: humans and baboons.[2] It would have been illegal and immoral to subject either species in captivity to extreme thirst for observation, but in the wild ecologists could study behavior of our nearest dryland relative, baboons, as an isolated troop confronted the end of water, and it was eerie and unnerving to watch.[3]

As one old and deep crocodile-infested pool evaporated, baboons adapted in ways remarkably similar to arid-adapted humans such as those bands in the Kalahari. Initially the troop kept its integrity, bonding across kinship lines by grooming. Males and females maintained their water balance by eating moisture-rich fruits. Soon the families dispersed in order to reduce impact on resources. The baboons dug and drank from nearby pits, kept to the shade, and remained subdued during the heat, conserving their energy while avoiding needless risks. But as drought wore on, as those pits caked into mud and then dust, these primates crossed some invisible line.

Casual cooperation broke down. Large males commandeered remaining puddles, and jealously guarded the slow seepage. Others began screaming at each other, slapping, hitting, clawing hair, and biting ears in the unrelenting heat. Inevitably, competition turned deadly. In 120-degree heat, thirty weak baboons collapsed from water deprivation. Dominant males emerged with bloody hands and teeth, obtaining moisture from the eyes and liquid parts of former colleagues. In order to survive, the primates had begun eating, and drinking, their own.

While humans had evolved far above such simian savagery, basic survival instincts remained locked within, waiting until the rains failed several years in a row. Whereupon protracted drought turned peaceful neighbors into hostile rivals who snarled fiercely at one another. In many dry regions of the world, previously congenial rural farmers, herders, and fishers had all, at some indefinable point, picked up their shovels and machetes and hacked each other over competing claims to vanishing creeks.

Like baboons at that shrinking water hole, *Homo sapiens* fought over water in mostly isolated sporadic violence, the kind of temporary insanity that appeared to have occurred, for example, in drought-wracked Jamestown Colony. In the name of securing resources, various stressed and thirsty individuals—albeit predominantly males—might commit assault, battery, rape, and murder if they could get away with it. But as society grew populous and complex, the nature of violence changed; hierarchies established loyal units for team defense and calm, tactical, organized aggression.

These units typically overlapped, like concentric circles. Americans bound allegiance to neighbor, city, state, country, NATO alliances, and the UN. At each level we united with others in a mutually beneficial relationship to form a tight-knit solidarity, securing and forcibly defending our shared interests in stability with security guards, sheriffs, SWAT teams, militias, armies, divisions, and peacekeeping forces, respectively. But as populations grew and drought bore down, thirst turned people against one another. The shortage of water resources threatened to undermine the bonds of what had been socially cohesive units. In Botswana, water-rationed suburban families snitched on local neighbors to utilities, which fought agricultural interests for a greater share of reservoirs. In the United States, cities and states gazed covetously at flowing water and mobilized quickly and aggressively to seize neighboring rivers as their own.[4] Around the arid world, competing jurisdictions divided the loyalty of police and militias and armies, as states seized watery borders, threatening war over the streams they shared.

STATE-SPONSORED WATER CONFLICT can be traced back to ancient Sumer. There, the city-state of Umma breached canals in order to defeat its rival Girsu's ability to feed itself.[5] Some 4,500 years later, the regional names had changed but the methods remained. To thwart hostile Shia opposition after the first Gulf War, Saddam Hussein diked, diverted, and drained off the vast intricate Mesopotamian waterways, destroying Earth's oldest wetland civilization: the legendary Marsh Arabs. New Turkish dams cut off water to drought-prone states, bringing tanks to its borders. Downstream, Iraq and Syria demanded their share of Euphrates currents, but Turkey claimed the bulk of rivers on behalf of the food and energy needs of 22 million citizens, who, as a consequence, saw water as a by-product of power. "If Turkey is stronger," an upstream farmer explained, "I get to keep my water."[6]

That farmer encapsulated what might be called the realpolitik of water, and it echoed in Middle Eastern power corridors. In 1979, Egyptian president Anwar Sadat darkened an otherwise festive Middle East peace treaty by announcing that "the only thing that could take Egypt to war again is water."[7] Jordan's King Hussein later pronounced that he, too, would never go to war with Israel again, which brought sighs of relief until he added: "except over water." The drumbeat continued to escalate. In 1988, Boutros Boutros-Ghali, later the UN secretary-general, predicted that the next war in the Middle East would be fought over the waters of the Nile, not politics.[8] Seven years later the World Bank's vice president told us to not worry about oil, since "the wars of the next century will be over water."[9]

America's post-cold-war national security establishment appears to have absorbed this assumption too, awakened in February 1994 by a haunting essay in the *Atlantic* magazine appropriately titled "The Coming

Anarchy," in which renowned journalist Robert Kaplan indelibly linked warfare with environmental shortages. "Nature has become a hostile force," he explained. Whereas, "democracy is problematic; scarcity is more certain," and turns previously friendly tribes, cities, and armies against one another as each vies to secure finite resources.[10] There were many scarcities to fear. But of all the diverse environmental security threats, one vanishing resource troubles analysts most, since all regimes collapsed without it: water.

As far back as 1871, Mark Twain quipped how, "whiskey's for drinkin' and water's for fightin' over," but in recent years, no one was laughing. North and South Korea raised tensions over rivers that crossed the DMZ, while the two newest members of the nuclear club, India and Pakistan, warned each other of dire consequences that could result from damming the shared Indus River. Alarm might be expected from the National Intelligence Council or *Foreign Policy* magazine, but even Nobel Peace Prize laureates from Mikhail Gorbachev to Al Gore warned that the world sat precariously on a tipping point of "water wars."

That alliterative phrase has been loosely interpreted. Many use it as colloquial shorthand to describe peaceful disputes between thirsty political interest groups.[11] Others more rigorously restrict the scope of the term to the violent use of water as weapon, tool, target, and technique.[12] Cases abound of armies dynamiting flood control dikes,[13] bombing hydroelectric dams,[14] targeting water treatment plants,[15] or using corpses to poison wells.[16] While horrific, even these examples deployed water as a tactical means to an end. But the narrowest definition—the one that obsessed national security experts as the climate changed—was a classic war between two sovereign states, strategically waged to secure water itself.

While nations could legally justify sending armies into war over water,[17] the good news is that, so far, none has. An exhaustive, landmark study led by Aaron Wolf at the University of Oregon scoured the last five thousand years and conclusively found no evidence to support classic water wars. To be sure, water scarcity always and invariably exacerbated international relations, in what theorists call *proschemata* and laymen call fearmongering, saber-rattling, and jingoism. But Wolf's research showed how even hostile neighbors—India versus Pakistan, North versus South Korea, Iraq versus Turkey, Israel versus Jordan—resolved water conflicts diplomatically even as disputes raged over other issues, and even while the disputants effectively ran out of water.[18]

Hawks conceded that classic water wars had not erupted yet. But, they argued, past was not prologue: There had been no oil wars before the marginal benefits of invading a well field outweighed the marginal costs of risking blood and treasure. Even the most upbeat water optimist predicted that "tension over water will make up yet another element in a potentially explosive cocktail of international conflicts of interest."[19] With enough thirst, drought, and desperation, water wars will surely come.

Tensions had risen even on North America's borders. Montana farmers once tried to dam the St. Mary's River that flowed north across it, provoking Canada to vow it would drain the Milk and divert the Columbia before either could cross south. Likewise, the United States and Mexico fought endlessly over the Rio Grande and the Colorado River—as both vanished under explosive growth that sucked down rivers and aquifers—until America couldn't deliver water promised to Mexico down the Colorado, while Mexico hit the wall on the Rio Grande. By 2004 the shared buffer of Falcon Reservoir had vanished, and stress was reportedly tearing apart a sixty-year-old international water treaty, "along with any pretense that the US and Mexico are not on the verge of political war."[20] The United States carried a big stick behind an extreme position drafted by former attorney general Judson Harmon, whose eponymous "Doctrine" warned Mexico that any rain that fell in, gathered on, was dammed by or diverted for America was America's—America's alone—and by sovereign status thus deniable to anyone but Americans.[21] Reminded that the hotly disputed Columbia River rose in Canada, America repudiated "the Harmon Doctrine, which is not part of international law,"[22] except, of course, on rivers flowing to Mexico.

Other transnational watersheds faced more stress, and fewer peaceful options. Geopolitical analysts discovered that 2 billion lives depended on water from 263 rivers, which crossed the borders of 145 nations.[23] Of these, they focused on likely flashpoints between thirsty states, calibrating relative stress factors that ratcheted up tensions and the odds that scarcity would cross that indefinable threshold of primates until armed hostilities broke out into wars waged over water.[24]

Certain states were ruled out: island countries, rainy countries, dry but oil-rich countries that could afford to desalinate the sea. Upstream hegemonic states like the United States and China were powerful enough to control tributaries, and downstream vulnerable states like Mexico or Laos, respectively, were too weak to do much about it. The analysis often boiled down to Africa, where a recent study of environment and warfare reached unsettling conclusions. In Kenya's dry season, Masai and Kikuyu began killing each other over control of the Weaso Kedong River, the only water source for thirty thousand Masai and their cattle.[25] Across the Dark Continent, the study compared rainfall levels and incidents of civil conflict; regardless of whether the country was well or poorly governed, the data found that as rainfall declined, conflict rose, with a statistical certainty of 95 percent.[26] So where in Africa might competition for water resources become most violent? Hypothetically, the most likely candidate would be dry, thirsty, landlocked, well armed, positioned both down- and upstream, utterly dependent on water, lacking a river or aquifer exclusively its own, and facing testy arid neighbors while trapped in a volatile, blood-soaked, war-torn region where 80 million people surge across borders chasing rain that refuses to fall.

BOTSWANA'S BORDERS WERE fragile and abstract, originally imposed by ignorant diplomats who never set foot in Africa and who cared little for its people. The Scramble for Africa carved through 190 culture groups, amalgamating ten thousand tribal monarchies and chiefdoms into forty colonies and protectorates like Botswana. "We have been giving away mountains and rivers and lakes to each other," groaned Lord Salisbury, "only hindered by the small impediment that we never knew exactly where they were."[27]

Coming from rainy Europe, these diplomats genuinely expected river borders to keep the peace, protecting precious land behind safe defensive moats. Just as Germany's Rhine formed a well-defined and solid barrier to aggressors, so they drew dotted lines based on vague reports of watercourses, and initially set borders right up to the stream banks, but no farther. In this fashion the artificial state of Botswana became little more than a vast imperial buffer zone to keep Germans, Portuguese, and Dutch Boers from encroaching and advancing on British claims. It was to be defended by nine moatlike rivers: the Nossob and Molopo; the Shashe and Limpopo; the Kwando, Linyanti, and Chobe; the Zambezi and Okavango.

Africa's harsh aridity reversed the rules of the game: Wet borders became targets, while dry land deterred. On tributaries forming Botswana's borders, the queen's diplomats shrewdly rewrote the precise border demarcation to include the entire current, thus denying rivals access to water and creating an inhospitable buffer zone of drought.[28] It was the first case where Botswana's leaders used water as a weapon.

Watery borders looked permanent on paper, but not long after ink on maps began to dry, so did actual rivers. Over the last century southern Africa's riparian borders were slowly erased by thirst, wind, sand and sun. Climate models showed the continent getting "severely short of water in decades ahead . . . particularly where rivers cross borders."[29] As those transboundary rivers began to vanish, Botswana's neighbors muscled in to divert off, dam up, or pump out what little waters remained. Southern Africa's evaporating borders brought the very aggression they had been initially designed to deter.

Paradoxically, the region's international tensions rose directly from the spread of domestic peace.[30] During their neighbors' bloody civil war decades in the late twentieth century, the tranquil Botswana had enjoyed unfettered access to soak up its wet borders, becoming the continent's richest per-capita country in the process. Then peace broke out in the region at the worst possible time.

Peace drives development, which requires water, which demands governments to claim their fair share from border-crossing rivers. Postapartheid South Africa now sought half the Limpopo River,[31] upstream Zimbabwe grabbed the Shashe River, and Zambia prepared to dam the Zambezi. Angola could soak up 98 percent of the Okavango, which originated in its country,[32] while Namibia might divert and harness whatever was left for its own thirsty capital.[33] Every drop a rival withdrew deprived Botswana of its lifeblood, and soon the country's usually peaceful citizens and leaders spoke openly about killing and dying for water.

Botswana's government viewed certain rival water plans as acts of aggression and escalated military spending. The country purchased a dozen fighter-bombers, tanks, and medium-range artillery.[34] The Botswana Defense Force mounted troops along rivers in unlikely, sparsely populated places. It threatened rival Namibia in a hostile border dispute that hinged on the depth and course of a shared watercourse.[35] The long-awaited dream of a Pax Africana was becoming a dry descent into hell. And at the center of this geopolitical drama lay a delicious irony. As rival states moved in on its watery borders, President Mogae found himself occupying the uneasy position he had forced upon the Bushmen: landlocked, isolated, thirsty, and facing a hot dry future. The besieger had become the besieged.

CHAPTER 19

Intimations of Genocide

LONG AFTER QOROXLOO LOST GOVERNMENT water, the men with guns kept pouring in. Year after year Botswana convoys closed in on her with increasing frequency; the rare trickle became a regular stream of recognizable faces. Some officials arrested suspected hunters. Others reminded the thirsty dissidents about all that water and all their families waiting just outside the Kalahari reserve. Most flexed government muscle, taunted Bushmen, and made mocking threats to drive them out.

Blocked out by the siege, friendly visits weren't nearly as frequent. But every so often Qoroxloo's supporters and I could sneak in via tourist vehicles and smuggle in the latest news of the international fight raging beyond the reserve. One day a hostile convoy coincided with our friendly visit.

I had brought FPK's Roy Sesana and WIMSA's Mathambo Ngakaeja to Metsiamenong to meet with Qoroxloo, Mongwegi, Nyare Bapalo, and other dissidents inside. All parties shared what they knew, and each asked for advice, plotting and scheming tactics in an open forum until they could agree upon a shared strategy. But thirty minutes into their debriefing, all conversations stopped. Some Bushmen rose to their feet, listening to the all-too-familiar noise of government Land Cruiser engines coming from the south. Based on previous run-ins, Qoroxloo and Bapalo had strong suspicions of their motives. Earlier, Bapalo told me, "[Minister of Local Government Margaret] Nasha predicted that we [who remained] would be tortured by thirst and hunger and would ask to be relocated by the government. But I didn't ask as they predicted. Later, I was surprised that the district commissioner came here all the way from Molepolole to ask me to relocate. Again I stayed. But he promised that the government would return, and it was going to bring more people the next time. So maybe these are the people he meant, and they have come here to fight."

Sesana and Ngakaeja had been tracked by officials after they bought water and provisions at a store. Days earlier, while the two leaders were addressing Bushmen moved outside the reserve, a police truck had circled, watched, asked a few questions, and departed in a hurry. By traveling with foreign tourists like me, Bushmen couldn't be detained without risking a diplomatic incident. But at the reserve gate we overheard an official tell Sesana, "I will follow you to the ends of the earth."

Now the government vehicles plowed right into the settlement, and several men dressed in camouflage fatigues dismounted. Two swaggered casually toward our half circle squatting on the sand.

"Maybe," whispered Bapalo, rising at the officials' approach, "maybe they have come at the district commissioner's request, sent with guns and sticks to 'kgatho' me."

Kgatho, an interesting word choice, derives from a hunter-gatherer's technique; it roughly translates as "to forcibly extract," say, a water-embedded plant or scrub hare or steenbok from its Kalahari habitat. Caught out in the open and exposed, these species lost their individual identity and structure and were soon broken down. But *kgatho* also has a political equivalent: to eradicate the identity and structure of people through amoral force, using water; to extinguish by dehydration.

To prevent that, Bushmen and their supporters pushed back, deploying a word equated with evil. When it cut off water to the central Kalahari, human rights activists alleged, Botswana's government was "writing the latest chapter in the genocide of the Bushman peoples of southern Africa." The feminist icon Gloria Steinem took up this accusation, citing the Kalahari siege, where "cultures of enormous sophistication and importance to everyone in the world . . . are being exterminated. This is, in fact, cultural genocide, as these scholars of cultural genocide have documented in a case that's being brought before the International Criminal Court."[1]

There was no quantitative comparison with the horrors of Dachau or Rwanda or Cambodia's Killing Fields. Neighboring Zimbabwe's Shona rulers killed more Matabele minorities in one day than all Bushmen who died unnaturally over a decade in Botswana. So to avoid overstating the threat, yet still use the "G" word, most critics inserted qualifiers. Botswana's dehydration tactics were thus "the last chapter of ongoing genocide," "verging on genocide," "tantamount to genocide," "linguistic genocide," "economic genocide," or "cultural genocide."

Perhaps such agonized modifiers may not have been necessary. Speed and numbers alone do not define genocide; the real issue is intent and motive. The goal of invoking genocide is less to pass self-righteous judgment than to halt atrocities before a body count can accumulate. The Polish-Jewish scholar Raphael Lemkin coined the word not to punish past nationalistic "crimes of barbarity," but rather to stop their recurrence. Under Lempkin's 1943 definition:

> Genocide does not necessarily mean the immediate destruction of a nation . . . It is intended rather to signify a coordinated plan of different actions aiming at the destruction of essential foundations of the life of national groups . . . the disintegration of the political and social institutions, of culture, language, national feelings, religion, and the economic existence of national groups, and the destruction of the personal security, liberty, health, dignity, and even the lives of the individuals belonging to such groups.[2]

This was old news to Bushmen. Exactly eight decades before the word *genocide* was coined, British observers reported "a wholesale system of extermination of the Bushman people" of Africa's Cape Province, and urged intervention with land set aside to protect the remaining survivors.[3] At the dawn of the twentieth century in the German colony that is today Namibia it became official policy to eliminate Kalahari Bushmen, who were differentiated on genetic grounds and because they could never be assimilated.[4] After the First World War victorious British-Boer troops in the region carried on under instructions "to flush out the Bushmen and to destroy the hordes . . . None of them remained alive: neither man nor woman nor child were spared."[5] All this might sound eerily familiar; modern political theorists traced the roots of European totalitarianism and forced labor camps to the colonial experience in Africa, specifically to the Kalahari's stateless ethnic and racial indigenous minorities.[6]

But all that took place before Lemkin's definition. To emphasize preventive rather than reactive measures, consider the warning signs on genocide's precondition checklist.[7] Members of the dominant governing society (Tswana) had to have a strong centralized authority and bureaucratic organization. Check. They needed a sense of superiority among the majority. Check. Those in power must conversely perceive their potential victims as less than fully human: as dirty, crude, pagan, savage, uncouth, barbarian, degenerate, lazy, outlaw, ignorant, racially inferior, prehistoric creatures. Check. Genocide involved "deliberately inflicting on the group conditions of life" any "bodily or mental harm" calculated to "bring about the physical destruction of culture and connection to habitat." Check. Its targets could include "in whole or in part" a minority of the minority, like one thousand Kalahari residents out of fifty thousand Bushmen, out of 1.6 million citizens. Check.

There was one additional precondition that analysts have found worthy of taking into account. Perhaps the most volatile element with the potential to tip repressed racial hatred into unchecked genocide was natural resource scarcity.

To be sure, scarcity alone may not trigger the spasm or cascade of horrific events, and root causes of genocide weave back through complex strands of ethnic, cultural, rhetorical, historical, and psychological tensions. But in Rwanda, and elsewhere, some analysts return to one underlying ecological reality: A decade of stressed land and a dwindling water resources base could no longer support the nation's subsistence population.[8] By 2008, this reality was better, or at least more prominently, appreciated. "Too often, where we need water we find guns instead," said UN secretary-general Ban Ki-moon, who observed how in western Sudan, "fighting broke out between farmers and herders after the rains failed and water became scarce." In the ensuing genocide, two hundred thousand have died. Several million have fled their homes. "But almost forgotten," he added, was "the event that touched it off—drought. A shortage of life's vital resource."[9]

SOUTHERN AFRICA'S VITAL resource was growing so short that the regional drought seemed likely to touch off a similarly violent event. The five best-armed countries in the region already suffered less than an average fifteen inches of rainfall; another expected 10 percent drop would reduce trickles a further 70–80 percent. Angola was grasping at the Okavango headwaters, famine-struck Zimbabwe was trying to replug its dams, South Africa's reservoirs were daily evaporating fourteen gallons per person, and Namibia's cities lost twice what they required to the unforgiving sun.[10]

But those observers watching borders for the outbreak of water wars were looking in the wrong place. Caught in the middle, closed in on every side, President Mogae quickly deduced that Botswana could hardly risk a water war against several well-armed and sovereign nation-states, and so instead his government rushed, with quiet desperation, to negotiate the best terms he could get. Talking fast, his generals sat down with their counterparts among the equally thirsty, border-encroaching rival states and began horse trading. His politicians set up representative river basin forums on the Orange, the Zambezi, the Limpopo, and, most urgently, the Okavango. His diplomats raced to revise and endorse the Southern African Development Community's Protocol on Shared Watercourses, then bargained hard over water allocations. To be sure, Botswana compromised with each rival, but in doing so held at bay the risk of regional violence. The president's representatives discovered the state could save more blood and treasure through exchanging of water's benefits—virtual water embedded in food, energy, tourism, cattle—than it could killing over the real wet stuff. "The water wars that the popular media would have us believe to be inevitable," quipped Tony Turton, a regional hydropolitical expert, "will not be fought in the battlefield between opposing armies, but on the trading floors of the world grain markets between virtual water warriors in the form of commodity traders."

Since Botswana could trade what it controlled, this was correct. But it didn't help the Bushmen trapped in the Kalahari. As Mogae peacefully negotiated over vanishing transboundary waters, his state regarded more dangerous rivals to be those disloyal groups competing for water within its own borders.[11] And so it turned its guns not on foreign enemies but upon those occupants of the nation's arid heartland.

This follows the unspoken rule of what may be considered unequal or "asymmetric" conflicts, in which sovereign nations fight stateless people over water. Israel made peace with Jordan but seized the Palestinians' West Bank in part to secure its three vital aquifers.[12] Iraq, Syria, and Turkey would not wage war with each other over the transboundary Tigris and Euphrates, but all three killed Kurds who lived atop the region's water sources.[13] For water security reasons China has not fought sovereign Myanmar or India but did absorb Tibet, the provincial source of two thirds of Asia's waters;[14] Sudan glommed onto the internal supplies of water resources in Darfur;[15] and while the United States never went to war with sovereign Canada or Mexico over the transboundary Colorado, Columbia, or Rio Grande, each of those nations has confiscated aquifers and rivers long held by Native Americans, acts invariably carried out in the name of locking up national security for "the greater common good."[16]

Likewise, President Mogae lacked the money or firepower to, say, invade Namibia and Angola, capture the entire Okavango River headwaters, depopulate the watershed, and brace for massive international retaliation. Yet he had done exactly that within the Kalahari reserve, where control of water had become both a tactical means and a strategic end. So even while retreating from open conflict against surrounding neighbors, Botswana's military leaders intensified pressure on those indigenous citizens trapped at its core. The state concentrated the bulk of its energy inward, setting sights on those deemed to be the state's most dangerous foe, closing in on Qoroxloo's band, who had no weapons with which to fight, no rights from which to negotiate, and no place left to run.

To be sure, Botswana did not gas or annihilate its Bushmen, nor did it round Bushmen up by the hundreds and bind them and embark on their wholesale extermination, to be buried beneath mass graves. By current definitions, it didn't need to. Botswana merely had deliberately and forcibly to shut off water, concentrate Bushmen elsewhere, spread disease among women, separate children from parents, burn houses, sever access to food, and ban traditions all in order to assimilate the last remnants of an endemic national group. Moved by the water cutoffs and siege, the British writer Sandy Gall shed his earlier modifiers: "It is genocide, alright, but not by stealth. It is open, unashamed, and contemptuous of world opinion."[17]

Perhaps that's why each Kalahari confrontation was regarded as a "last stand" in the popular and legal imagination. Both sides knew the stakes escalated each time the government closed in, each time Botswana tried to *kgatho* them. And yet, a careless courage can emerge among people pushed too far. Stripped of their rights and their families, deprived of government water, the remaining unarmed Kalahari dissidents had nothing left to lose and pushed back.

The armed officials walked slowly toward Nyare Bapalo, but before they could open their mouths, he spoke first. His voice was not gentle, or timid, or vague. He angrily told them to go away. "Stand aside, we are still in our meeting here."

Taken aback, the officials stopped. They looked surprised, even hurt.

"Even if you have your guns, I'm asking you to stand aside."

Emboldened, Mongwegi, a talkative mouth-harp player, chimed in as well. "You are not invited. We are having a meeting, and you just barge in on us."

The officials glanced at my colleague and me, two white foreigners holding cameras. "You misunderstand," they said, all smiles. "We are simply coming to . . . to greet you!"

Said Bapalo, "That does not matter. You are not welcome. You can't divorce from your wife and then come back and expect to be welcome. You must go and collect many people if you want to remarry. You need official approval."

"Your animals are out there, outside and beyond our village," said Mongwegi. "Go to them. Go look to your animals, not here."

Soon the others in our meeting, including Qoroxloo, began chiming in, standing and pointing and shouting at the government officials. It all seemed very un-Bushmanlike, and quite effective. In less than two minutes the officials had retreated, turned back to the vehicles, and drove off.

The unarmed "harmless people" who refused to be *kgatho'd* from the CKGR had instead evicted rivals from their ancestral home. Ngakaeja had watched the entire altercation in silence. He had come hoping to inspire the Bushmen inside the reserve, only to find it was they who had rejuvenated his own spirits. "These people out here," he repeated quietly, "they are strong."

As the men in khaki receded beyond the horizon. Bapalo urged them on: "We told you we're not moving. Bring your big guns next time 'cause it is only through our death you will win."

CHAPTER 20

Escalation of Terrorist Activity

IF RESOURCE SCARCITY NUDGED GOVERNMENTS closer to genocide from above, dehydration might lead stateless people to opt for terror from below. Indeed, that desperate act may even be seen as a perfectly reasonable response by anyone so deprived: Attack those people and institutions who deny you water. It is a testament to their self-control that despite watching their sons and daughters suffer unnecessary thirst, hunger, and physical abuse, the Bushmen never resorted to violence; they raised voices, but never a fist, much less a poisoned arrow. Nevertheless, dissident Bushmen were still considered a national security threat and their activities were carefully monitored throughout the legal proceedings, just in case tempers boiled over. And while throughout the ordeal over access to water President Mogae and Qoroxloo Duxee never met face-to-face, their struggle was waged through the political maneuvers of two intermediaries who fought on their behalf: Mogae hired Sidney Pilane; Qoroxloo empowered Roy Sesana.

Sidney Pilane was born for politics and thrived in courtrooms. He was built like a running back, his robe fit perfectly, and when, after each session, he removed it with a flourish, a dapper tailored suit was revealed beneath. Ambitious, daring, abrasive, and unyielding, he ran a lucrative criminal law firm until 1999, when the exuberant young legal genius rescued President Mogae from an electoral scandal. From that moment, Pilane became the president's own Karl Rove. As "special adviser," Pilane had a vague portfolio and around-the-clock access to Mogae, who handpicked him to represent the country against the Bushmen menace.[1]

His political counterpart could hardly appear more different. In contrast to tailored shirts, Roy Sesana wore: traditional beads; the skins, head, and horns from his totem Steenbok; dark sunglasses; an Adidas sweatshirt; and pressed cheap slacks. Outside of court, the dapper Pilane pointed to this motley outfit and smiled. He started to laugh, and others began laughing along with him. "You see here, you no longer wear only skins of Basarwa, but Western clothes. Don't you want to develop like the rest of us?"

"Yes," Sesana responded evenly, never taking his eyes off Pilane. "We want schools and rights and goods and services like water. But we see no reason why we can't have them where we are. Where we live."

Pilane shook his head in protest, looking about him for support.

Sesana continued, smiling, "You did not leave your home for services, did you? Nor will we."

"But that reserve is a place for wild animals!"

"So was the place where you were born and live. All of Africa was once a place of wild animals. Now there are none around here, but where we live there are."

His rhetorical judo left reporters wondering whether 99 percent of Botswana was living the wrong way in the wrong place, and not these Kalahari misfits. The appeals court gave the day to the Bushmen, recognizing the urgency of their case to go forward, in a ruling that encouraged Bushmen attorneys but not Roy Sesana.

That morning he had said, "[I feel] strong, like a hunter on behalf of my people. Sometimes we make a kill. Last time we did not. But whether this time we kill or not we will go home with our people." Now, forced to wait for a verdict, Sesana had to explain that he had neither made a kill nor missed his quarry but simply let fly an arrow that had yet to land. His attorneys predicted it could take months; for him that was too long. They urged tolerance. He grew impatient. He banged his fist on a table, vowing to lead exiled Bushmen back home, like a modern Gandhi-of-the-Bushmen, "back right straight through the gates even if we are stopped, beaten, or shot."

That was in early 2002. Three and a half years later, after years of crushing drought, the arrow still had not landed.

By then, Sidney Pilane was growing increasingly angry and somewhat erratic; the court case appeared to be taking a psychological and physical toll. First he collapsed in court, only to suffer the further humiliation of being rescued and revived by Bushmen attorneys. Outside the court Pilane got caught with his pants down—fleeing a jealous husband in an embarrassing incident that made the news—and began to make uncharacteristic tactical slips. He persuaded the ruling party in Parliament to alter and remove parts from the Botswana Constitution that he believed protected Bushmen rights, only to find the clause he had painstakingly scrapped at a high political price was, in fact, legally irrelevant. He had spent years convincing judges that diamonds had nothing to do with Bushmen relocation, only to have the court learn of a dramatic increase in mining concessions the very year the government cut off water. He played the race card, emphasizing the white skin color of his opposing attorney and Survival's director, but in the process underscored racial prejudice between Tswana and Bushmen. Finally, on September 1, 2005, Pilane lost it.

When the court ruled a certain report irrelevant, Pilane demanded to know why. He didn't like the explanation and demanded a recess "for personal reasons." When he came back, ten minutes late, the judges were displeased. So was Pilane, and as he complained, Justice Unity Dow reminded him that it was common courtesy when addressing the court to stand up.

Pilane refused, repeatedly, until Justice Dow calmly cited him for contempt and ordered bailiffs to seize him. But Pilane was not about to be arrested like a common criminal, humiliated in that courtroom. With his dignity to protect,[2] he fled the courtroom, burst out the doors, instructed his sidekick, Director of Wildlife Jan Broekhuis, to drive, and away they roared in a cloud of dust. After a few curt phone calls, President Mogae made Pilane drop his appeal, serve three days in prison, and offer an abject, unconditional apology. In court, Pilane groveled a bit, regretting "this unfortunate incident," affirming his deepest respect for the justices, showing public restraint, smiles, and subservience. But his impatience and frustration with the Bushmen case continued to simmer beneath the surface, and it was only a matter of time before his temper found an outlet for release.

Roy Sesana too had grown impatient with the courtroom, its stilted language and strange, slow ways. Justice delayed was justice denied. While waiting for a verdict, one out of ten of his Bushmen co-plaintiffs had already died, and just then, even more were being marched out at gunpoint. Right after Pilane fled the courtroom, perhaps not by coincidence, the government sealed off the reserve to outsiders. Inside, armed wildlife scouts, now joined by police, surrounded Qoroxloo's band. Patrols began to step up pressure tactics, firing over the heads of Bushmen, seizing possessions, issuing death threats, confiscating radios, arresting hunters. Intimidated Bushmen found it impossible to move about within their own homeland. His colleagues had been arrested trying to bring water into the reserve, but at least they had tried; Sesana, who had promised to lead the Bushmen back into the Kalahari himself, remained stuck in this eternal purgatory, sitting through legal tedium, unable to lift a finger. He decided the time had come to do more than mimic his tormentors during cigarette breaks. He would defy them, acting not only outside of the government's restrictions but also against his lawyer's advice.

The following week the Bushmen attorney Gordon Bennett warned the justices they were being bypassed. By "taking irreversible steps to remove from the reserve those people who were now living there," he said, the government was "anticipating the decision of the court."

How did Bennett obtain this information?

With great difficulty, he replied. The normal means were closed off; even he had been blocked from entering the reserve to speak with his clients because of the tight blockade. "So, it is a real problem, that they are in effect incarcerated within the reserve and out of contact with the outside world."[3]

The court duly noted the grave situation. But before it adjourned, it wondered aloud, Where was Roy Sesana? Bennett professed he had no idea. Pilane did. The government learned from its informants how Bushmen were planning to break the siege in broad daylight and bring water to those sealed inside the reserve. That attempt could not be allowed to succeed. If Sesana tried to carry in water, Pilane determined to stop him.

At nine in the morning of Saturday, September 24, a military helicopter landed on the soccer field at Ghanzi. The event generated much excitement. Dozens of children rushed out and saw Sidney Pilane and Jan Broekhuis on board. It flew over the Kalahari reserve, viewing people on the ground, then returned within the hour. Those inside boarded a government 4×4 and drove down the road toward New Xade, the government resettlement camp known to Bushmen as "a place of death."

There, a boisterous assembly led by Roy Sesana emerged from the northern side of New Xade. People were in an anxious mood. Vehicles honked, escorted by two hundred Bushmen singing Christian melodies but with words of a somewhat different spiritual pilgrimage: "Re a tsamaya re a gae CKGR" (We are marching home to the Central Kalahari Game Reserve). The Bushmen, who said they felt like freedom fighters, were followed closely by a Criminal Investigation Department car. Sesana appeared unfazed by the police presence; he expected some kind of altercation. To stiffen resolve, others bought drinks at local shebeen and said good-bye to New Xade, geographically and spiritually. Long-winded speeches grew melodramatic, but fear was palpable. Moruti Daoxo Xukuri, the chairperson of FPK, proclaimed, "Jaaka Morwa Modimo a lo swetse le nna ke a go lo swela" (Just like the Son of God died for you, I am going to die for you as well).

The crowd said, "Amen."

Sesana's assembly had packed five vehicles with containers filled with water and food, and drove slowly toward the central Kalahari.

Two miles down that road they met armed forces under the direction of Sidney Pilane.

The Bushmen stopped.

"What are you doing?" shouted the government's men.

"We are taking water to our families inside the reserve," they answered.

The Bushmen vehicles spread out. A few tried to turn around in the sand. As they did, facing perpendicular to the government, a dozen heard Pilane shouting out to the officers, "What are you doing? Can't you see what they are doing? Fire on them!"[4]

They fired. First, they shot tear gas into the crowd. Bushmen's eyes began watering; some vomited. The truck beds emptied and some drivers and passengers were brought to their knees.[5] The government fired again, this time spraying rubber-encased steel bullets into the crowd.[6] One penetrated Motswakgakala Gaoberekwe in the right side of his face, breaking his jawbone and sending him down writhing in the sand.

One police officer handcuffed Roy Sesana and then, under Pilane's instructions, struck Sesana's face, knocked him down, and stomped on his back. On Pilane's order, the entire group—men, women, and children, including a boy also struck by a rubber bullet—was arrested, handcuffed, and quite literally read the riot act.[7]

Pilane later explained he ordered the police to open fire because the Bushmen brandished knives, bows and arrows, spears, and clubs, preparing to fight the police. He produced no evidence whatsoever to substantiate his claim. The government was able to confiscate nothing from the Bushmen, no dangerous or deadly weapon, unless that category now includes water.

EVENTS SUCH AS THIS only helped fuel the international backlash. As their reputation and market share deteriorated under the hostile propaganda war, Botswana and De Beers could have simply called off the siege. Instead they did what any affluent, embattled, and misunderstood commercial or governmental entity would do. They hired Hill & Knowlton.

As one of the world's five largest public relations and lobbying firms, Hill & Knowlton operated seventy-one offices in forty countries, including nineteen in the United States, with close ties to both Bush administrations. Its clients included Big Tobacco, BCCI, Enron, Walmart, and the (pro)-Asbestos Council. By hiring the PR services, Botswana joined the ranks of Idi Amin, the emir of Kuwait, the CIA, and the Chinese government shortly after the Tiananmen Square massacre. The firm employed 1,100 people and charged Botswana $350 per hour to transform "the Harmless People" into a dangerous threat, thus earning global sympathy for Botswana, customers for De Beers, and enmity for nonprofit activists such as Survival International and First People of the Kalahari.

H&K lived up to its reputation and earned its hefty fees. Within months of signing a contract, the PR firm reportedly helped arrange political junkets to the country while installing its Kalahari campaign handler into the London offices of the Botswana High Commission, at an apparent cost to taxpayers of $100,000 a year for three years. Finally, in the asymmetric water war propaganda helped ensure Qoroxloo's band was seen as part of an international network engaging in acts of terrorism.

Links between terror and water cannot be dismissed outright. After all, terrorists fought occupying forces by targeting a government's physical incarnation, including its water infrastructure.[8] On the very day Botswana cut off Bushmen water in the Kalahari, the FBI found a computer, linked to Osama bin Laden, that contained engineering software related to dams; al-Qaeda had been studying American waterworks in preparation for further attacks. The terrorist group specifically sought information on water supply and wastewater management practices in the United States and abroad.[9] President Bush identified water supply systems as one of eight vulnerabilities and urged Americans in cities to report suspicious behavior, "watching over key utility assets such as water tanks, reservoirs and even fire hydrants."[10]

Indeed, following 9/11, security analysts warned that $7 billion to analyze WMD attacks was being misspent, since the United States ignored or shortchanged the risks from a fertilizer-stuffed Ryder truck parked atop a dam wall. Terrorism experts distinguished between water deployed as tools (poisoned reservoirs) or as targets (dams dynamited with cities downstream), but had ample precedents.[11] Even before World War II, FBI director J. Edgar Hoover warned that America's "water supply facilities offer a particularly vulnerable point of attack."[12]

In terms of motive, some of America's first domestic terrorists were driven to violence by the loss of water. When the Metropolitan Water District secretly bought up all the water rights to the Owens River in California, farmers and ranchers blew up diversions in a futile effort to keep their waters from flowing off to Los Angeles toilets, lawns, and swimming pools.[13] In 1934, Idaho fishermen blew up Sunbeam Dam on the Salmon River, which blocked formerly prodigious spawning runs.[14] In July 1999, engineers discovered a homemade bomb lodged in a dam outside South Africa's capital; three years later in that same country, several white Afrikaaner suspects were tried for attempting to blow up the unmanned, unpatrolled Gariep Dam after taking a "Warrior Oath" and promising to unleash a "Night of Terror."[15]

But the worst, it seems, were those "terrorists" who got in the way of national water development plans, and Hill & Knowlton offered an instructive model in how to portray them and frame their activities. Through its careful lobbying and strategic public relations campaigns, the alcohol and tobacco industry has managed to frame all other addictive, deadly, abused substances as the real "drugs." Now Mogae's government officials—having destroyed and sabotaged water infrastructure in a calculated effort to create fear and psychological intimidation among civilian populations—tried to define Bushmen as the real "terrorists." The tactical ploy was audacious, but it caught on.

Major General Moeng Pheto—who led water cutoffs against Qoroxloo's band—pronounced criticism of his activities "a war against the country."[16] Government officials accused the Bushmen of attempting to create a breakaway state within the Kalahari reserve. To justify his actions in and out of the court, the state's lead attorney, Sidney Pilane, often casually referred to people like Qoroxloo as terrorists.

That domestic "terrorist organization," FPK, was a shifty one, but I managed to track down its ringleaders in a barren cinderblock room, where they had received a half-dozen death threats and were constantly running out of cell phone credits and electricity. Its chief, Roy Sesana, tried to borrow gasoline money, and hitched rides on numerous occasions to this makeshift office equipped with one chair and a landline that didn't work. FPK seemed to run on the fumes of nostalgia, for only when I asked about its origins did Sesana's face light up. He walked over to an empty filing cabinet, picked up the office's only photograph, and pointed out to me the seven men standing shoulder to shoulder. On the far right of the picture stood a Bushmen elder named Khomtsa Khomtsa. Third from the left and dressed in white stood a charismatic man—the mixed-blood son of a British rancher and Nharo mother—named John Hardbattle. A decade earlier Khomtsa had approached Hardbattle in the Kalahari and asked, "Whose land is this?"

Hardbattle had replied that all of the "deep sand places" belonged to the Noakhwe (red people) "for they have never sold it."

"You are a man of two worlds," said Khomtsa. "You can sit at the fire of our mother's people, and you can get up and sit at the table of your father's people. Will you show us the path, will you open the doors for us, so our voices can be heard?"

Hardbattle had answered by forming FPK, and with Sesana set out to make Bushmen's voices heard, starting with an international conference on sustainable rural development. There, Hardbattle observed how "twenty-six years of independence had brought Botswana forward and us, the First People of the Kalahari, backward." In one sentence Hardbattle committed three unforgivable sins: He distinguished tribes where Botswana pretended all were one; he acknowledged Bushmen as the first inhabitants before all later arrivals; and he criticized a proud government in front of foreigners. That simply was not done. The government tried to isolate this rogue cell before it could grow. It tried to ban FPK travel or contact with reporters and foreigners,[17] efforts which backfired. FPK won glowing profiles in the *Washington Post* and the *Times* of London, sympathy and resources from Prince Charles and the UN Human Rights Commission, debate in the House of Lords and U.S. Congress, and strategic alliances with Native Americans. Led by Senator Patrick Leahy—and through him the USAID, UNDP, World Bank, and other powerful entities—America sought reassurances that Bushmen did not suffer the discrimination and hardships indigenous people in our own country had. Botswana assured the United States it had no plans to forcibly remove Bushmen from the central Kalahari, although it warned Ambassador Howard F. Jeter it might cut their access to adequate and reliable sources of water. By 2002, Leahy was out of power, Hardbattle and the Bushmen's international champion Laurens van der Post were dead, and Sesana was the only survivor in the photo. With the United States and United Kingdom enmeshed in their own war on terror, Botswana shut off Bushmen water forever.

A domestic terrorist cell can be isolated; it lives only by infusions from a global network. Initially, Sesana reached out to potential domestic allies—Kuru Development Trust, Ditshwanelo, the Botswana Society, and the Botswana Council of Churches—to help it resist the siege. But one by one they urged FPK to compromise. Give up its protests. Call off its boycotts. Settle quietly out of court.

Sesana refused, and dismissed these groups as having become financially captives of, and thus politically beholden to, the government and De Beers. To a certain extent they agreed. "We have to work here," one explained to me, by way of justification for refusing to fight. "We have to raise families and send our children to schools here." Nor could Sesana seek help from rival nation-states, which were now allied over water with Botswana.

But standing behind FPK's tactical moves, strategic alliances, Internet links, growing clout, legal counsel, and fund-raising capacity was Survival International, and soon that global network became the main target of Botswana's attacks. Dr. Akolang Tombale, permanent secretary in Botswana's Ministry of Mineral Resources and Water Affairs—and deputy chairman of De Beers in Botswana—accused Survival International of being a "terrorist organization."[18] His colleagues said Survival should be tried in The Hague's International Criminal Court. The rhetoric seeped down throughout the country, permeating all debate. Editorials condemned Survival and FPK for holding the country hostage. As Botswana's hardships lingered under the protracted drought, civilians joined the fray. In one influential article that captured the mood of many, Lesedinyana Odiseng, a doctor of occupational medicine, blasted the Bushmen's "unprovoked and immoral war against the Botswana Diamond Industry. [Survival] seeks to destroy the economy of Botswana to achieve its objective irrespective of the possible consequences of this on the lives of all Batswana. On this basis I suggest that SI is a terrorist organization and it should be treated as such."

I tried to treat it as such, but Survival also proved disarmingly easy to infiltrate. One could, without being stopped, walk in off a downtown London street into narrow offices and stride into the stifling cramped quarters of its mastermind, Stephen Corry.

Corry was a slender, bookish, earnest anthropologist with a pale complexion, unsmiling lips, and polite manners. His handshake did not intimidate, but his words often seemed to.[19] Corry told me he believed the Bushmen could not get a fair trial, let alone win their case, despite their popular support in the country, so with reluctance he chose to focus like a laser beam on the court of public opinion instead.[20] Recognizing that public perceptions and policy were increasingly formed on the Web, Survival used the Internet to shrewdly unite millions of diverse "us," filled with charismatic images and heartrending stories, against a unified, faceless "them." The Internet transcended borders and brought its own kind of sunlight. While remaining outside Botswana, Sesana and Corry waged their relentless and insistent PR campaign over the Internet, setting up multiple Web sites, collecting hundreds of thousands of signatures of support, and scorching Botswana's diamond and government leaders with heat coming from everywhere and nowhere, cementing their reputation in Botswana as an enemy of the state.

The Bushmen or their proxies seemed ubiquitous. On a global tour, Bushmen showed up on Malibu beaches and in Manhattan salons; they addressed parliaments in Switzerland, Germany, Italy, London, and Washington, D.C. They were celebrated in stories about Hopi Indians and rockers, movie stars and supermodels. By contrast, during Botswana's own PR tours to offset the Bushmen campaign and promote the concept of "Diamonds for Development," former president Masire had to fend off hostile questions from suburban retirees in California, while President Mogae was subjected to hecklers at Oxford.

Even at home the political hierarchy was inverted. After their arrest, the twenty-eight Bushmen who tried to bring water in to Qoroxloo and her band were hauled off and crammed into Ghanzi's small prison cells, where they remained for several days. Four claimed they were beaten while in custody. Beyond the deprivations, families said they mostly hated the density they were packed into, forced to urinate or defecate in front of their own children. Four days' captivity is a long time for anyone, but especially for Bushmen unfamiliar with confinement within four solid walls. But on the day of their release the Bushmen, organized under the "subversive and terrorist secessionist" movement First People of the Kalahari, discovered they had won Sweden's Right Livelihood Award, commonly known as the Alternative Nobel Prize.

If classic transnational water wars will be fought on the trading floors of commodity market exchanges, asymmetric water wars within borders will be won or lost in virtual battles that nevertheless result in physical casualties. They are already being waged via e-mails, blogs, Web sites, shareholder meetings, multilateral lending forums, news reports, letters to the editor, and above all, radio and television. In Botswana, the protracted battles to win the water came to a head in a global radio broadcast, pitch-perfect in the storyline that H&K had framed: by resisting the waterless siege, the reckless menacing Bushmen cells and their international terrorist allies were really only succeeding at destroying themselves.

Paul Kenyon, the producer of BBC's *Crossing Continents,* decided he knew what *genocide* meant—"it meant killing on a large scale"—and so the overheated water war of propaganda that was being dragged out by Qoroxloo's besieged band was actually "making matters worse" for everyone. Botswana's good government, peaceful military, decent citizens, and magnanimous diamond cartel—they were were all trembling in abject terror of Bushmen dissidents and their allies.[21]

"I'm afraid of them," admitted Bram LeRoux, who ran a pro-government, pro–De Beers religious charity to convert and assimilate Bushmen who had been dislocated outside the central Kalahari.

"Why are you afraid of them?" asked the BBC producer.

"I think there is a general fear amongst NGOs and everyone out here for Survival. We're afraid of them because you say something, the next day it's on a Web site, it's in a newspaper, and it's on wherever. And if you cry out about it, you're in big trouble because then you're an enemy of the [Bushmen]."

LeRoux objected to any negative characterization of Botswana's actions. "There is no genocide taking place; there is no ethnic cleansing taking place. There is . . . people are struggling for their rights."

"What kind of impact do you think it has when Survival International uses language like that?"

"Well, you know, if you go, if you go so much overboard, I think that the effect in Botswana is that it sounds like terrorism."

Gotcha.

There was just one problem with this Bushmen-as-terrorists storyline, and the BBC confronted Sidney Pilane on it: "Why was it that your government cut off the water to the [Bushmen]?"

"The water was cut off when everybody had left the game reserve except seventeen people," Pilane answered, untruthfully.

"What was your intention by cutting off the water?"

"It was unsustainable providing the services. Have you been to the CKGR? Because if you've been there you will know that it has to be one of the most wicked terrains anywhere in the world."

"But by cutting off the water, weren't you really just evicting them?"

"No. Most of them had left, only seventeen remained."

"But you were evicting the final seventeen."

"No. And they were not evicted. They stayed on."

"But you knew they couldn't survive without the water supply you had been providing."

"Sir, they have survived to this day. They still live there, without the water. They still live."

PART VII

Elimination

CHAPTER 21

An Open Heart

PRESIDENT MOGAE WAS NOBODY'S FOOL. He caught on to how people still sneaked into the Kalahari reserve. Sympathetic journalists managed to get around checkpoints, bring in supplies, and smuggle out stories that cast Botswana in a negative light.[1] By December 2004, Mogae's officials attempted to plug suspected leaks.

They tightened the reserve perimeter, closing gaps, checking credentials more carefully. Even so, looks could deceive. Americans who posed as big cat enthusiasts on safari could easily turn their cameras on people instead, and few guards could distinguish a legitimate vacationer from a Bushmen ally. So like U.S. airport security agents, Botswana cracked down on everyone indiscriminately, scaring even legitimate tourists, like Veronica Molosiwa.

Molosiwa planned a Christmas safari in the Kalahari. At the gate an official told her group, falsely, that their chosen site deep in the reserve no longer existed, and instructed them to camp near the Xaka water hole, but to go no farther. They complied. That night, two vehicles with eight tipsy wildlife scouts charged in to make sure, harassing the campers and demanding evidence that her group was not helping or visiting any Bushmen. They even checked vehicles for potential smuggled excess water.

Tourists found these officials rude and clumsy, while the president found them merely ineffective. Somehow, his vast Kalahari siege was so porous that even Bushmen from outside were continuing to flow back in. So in early 2005, Mogae decided he could not wait for a judicial ruling and, citing "administrative reasons," reinforced his blockade. Officials checked identity cards not only of Bushmen but of anyone seeking entry. Ultimately, the government sealed off the reserve to everyone. Tourists, journalists, lawyers, and researchers were turned away, with the new extreme measures causing outrage even among disinterested citizens like Molosiwa. "What would any other Batswana [citizen] say if they met a roadblock and were told they could not enter their home village? Within our government, surely there is someone who has a conscience. The path of violence, already begun, will only lead to further prolonged misery. Botswana was once a peaceful country. Why must we bear arms against our own citizens to force them off their own land?"[2]

As the siege tightened, formerly docile opposition parties began to ask similar questions.[3] High-court justices expressed their concerns that the executive branch was preempting their coming judgment about the very law being tested in their courtroom. But as the president simply said of Bushmen, "I cannot allow them to go back."

Indeed, facing a rising national and international chorus of criticism, enforcement of the siege grew even more entrenched.[4] Officials clammed up, and Botswana imposed a news blackout. Week after week, as the drought wore on and the Kalahari siege was framed as a national security issue, public fear mounted and the political pressure took its toll.[5] Roy Sesana and his would-be siege breakers were jailed in September. By October 3, 2005, officials had arrested another eight Bushmen trying to smuggle water in to their families. Over the next two weeks, armed police officers and wildlife scouts evicted the last Bushmen from Mothomelo, Molapo, and Gugamma. Only the dissidents in Metsiamenong still remained.

Those Bushmen driven out at gunpoint took their stories with them; piecing these together confirmed that the siege was no longer passive or limited to the reserve's peripheral entrance gates.[6] The Bushman Kangotla Kanyo said, "I came out of the reserve to tell the outside world that we are really suffering with thirst and hunger. The police are camped at our settlement and we are not allowed to gather or dig anything to eat. The policeman called Dibuile told us, 'If you die, the government doesn't care. This is the time to show you that you are nothing.' The police said that we have to suffer thirst and hunger so that we will leave the reserve."[7] Old Molatwe Mokalake said his family had been forced out of Molapo, for the second time, against their will, at gunpoint. When the government came with guns, he said, "it felt like war. We were not allowed to bring water from outside the reserve." His granddaughter said the "police came and told us they had permission to shoot us."[8] Just before Botswana jammed and confiscated the last donated radio transmitter, Bushmen outside the reserve received a broadcast from those still inside at Metsiamenong: "They're firing over our heads. They're beating us. We don't know if we can hang on."[9]

Then, static.

Many at Metsiamenong fled the gunshots, melting into the bush. When they returned they found the government encamped on the periphery of their homes, hovering, watching, and waiting for further orders. The siege had constricted.

Qoroxloo's family said she had tried to keep her spirits up by sticking to her routine. But her routine involved gathering moist food through increasingly long walks over great distances, and dispersal was now forbidden. She was effectively put under house arrest, incarcerated on her own land.

Bushmen faced no charges leveled against them other than hunting and gathering in the way of their ancestors. What made it surreal was the knowledge that her imprisonment was self-imposed. Qoroxloo was free to depart; she had only to say the word, a word that could not be retracted. If she allowed herself to be carried off, she could never return.

Her brave face grew transparent. Her son said she had never been afraid of any natural dangers or threats, like snakes or scorpions or lions. But now, others saw, she was as troubled as everyone else. The eyes gave her away.

Those remaining in Metsiamenong later described their own fear. They felt intimidated by the immediate presence of hostile strangers lurking so close. Bushmen could hear officials laughing, smell their rivals' meat cooking. They recognized the uniforms of the armed personnel units of antipoaching wildlife scouts and the paramilitary wing of the police, called the Special Support Group. Faces grew familiar but never welcome.

Her family said Qoroxloo seemed especially nervous about the government vehicles careening about. She had seen officials force her younger brother, Owa, to run in front of one of the vehicles for miles as a poaching suspect, and now others like those tormentors waited out there on the edge. Years earlier, when officials had cut off the water, at least they went away. When they had carted off people and dogs and goats and belongings, the government drove off too. All previous times Qoroxloo waited until the officials departed, passing through like a violent thunderstorm. But now their presence lingered.

For the first time, her options were circumscribed. Her endless spaces were closing inward, and the choices that she had enjoyed all her life narrowed down to two: She could die a spiritual death from boredom outside the reserve; or physically wither and starve by staying here in the camp.

Qoroxloo's band watched one another grow somber. Weeks without meat made everyone tense and restless. Some nursing women began to run out of milk, and Qoroxloo must have heard their babies crying in the night. Any food or fluid Qoroxloo put in her own mouth would deny her hungry and thirsty descendants. Cut off from the Kalahari, the government forced Qoroxloo's heart to compete for survival against her own flesh and blood. Held in one place, her family now became her rivals.

Toward the end of October, Qoroxloo's band noticed her behavior was growing erratic. Never morbid, she spoke about death. Never openly rebellious, she spoke of defying the government restrictions. One night as the flames leaped and sparks crackled and the smoke rose, Qoroxloo began to hum and then added words to the tune. Slowly she started to clap and rise to her feet, swaying slightly. One foot rose and fell, then the other. Right stamp, left stamp. Hands clapping. Eyes closed. Mouth open. A rhythmic song issuing forth until soon, deliberately and unmistakably, in the eyes of all the remaining Bushmen dissidents at Metsiamenong, Qoroxloo Duxee began to dance.

Each dance is as unique as the dancer, but in broad ways, dance stands as one of the unifying traits of the Kalahari Bushmen clans. And the "trance dance" stood out above all. Doctors and anthropologists probed the origins and meanings of dance to tap the blood root of human evolution. They cross-pollinated ethnobotany, rock art, oral history, psychoanalysis, and even testimony of hallucinogenic drug users and discovered many layered complexities of the dance. Dance went beyond a mating ritual of the young. Nor was it merely a celebration of a triumphant harvest, hunt, or wedding. In the Kalahari, Bushmen danced at the instigation of the elders, with less emphasis on displaying healthy, attractive individuals than on collective exhaustion to bring group harmony.

Dance arose naturally. Like hunting or gathering, it was never formally scheduled. It fused politics, religion, and medicine, all carried out in the course of events.[10] There was rarely one initiator; like political leadership, the call to dance circulated among the band, in this case: Qoroxloo. Nor was it mandatory; some joined only reluctantly, knowing the fatigue and collapse that might come through strenuous exertions. But that night, one by one, people in Metsiamenong joined Qoroxloo as she clapped a complex rhythmic beat and sang, wailed, or yodeled in her tremulous voice.

According to her children, Qoroxloo danced to relieve intense doubt, stress, and protracted discontent and to soothe frayed nerves as language grew sharp and tempers short. Her dance aimed at any general, nonspecific evil at work, a kind of exorcism. When people turned to her for help, she danced to heal.[11]

Qoroxloo's husband described her as "amazing when dancing, because she could dance with her hands in the air and a few moments later you see her swooping and bending downward with her hands wide open like a bird." She danced with an abandon unmatched even by women who were many years younger. Time had not slowed her. Friends and family members stopped worrying for Qoroxloo's health or safety and simply enjoyed the mood of anguish and joy that night as she started clapping, chanting, and swaying by the fire, rising amid the smoke, pounding her feet, swaying, and crying out, surrendering herself to the rhythm for hours and hours and hours until dawn. The dance over, her band was drained of energy, relaxed.

The next morning her children said she revealed an inner peace, "an open heart" matched by willful determination. She would defy the explicit commands of the armed patrol and cross their periphery. She would gather food and water.

Qoroxloo knew the risks. Men with guns warned they would kill Bushmen caught seeking food. Gathering was illegal. The only way to escape death, imprisonment, or expulsion was to avoid the vehicles that surrounded Metsiamenong. Overcoming her fear of them, Qoroxloo proposed to sneak past them in the morning, and while the group might find nothing, she said she had to try. Out there in the Kalahari lay wild foods, tubers, fruits, and roots for the children and babies, anything to keep them strong. Some tried to dissuade her, reminding her of the consequences, but she stayed resolute. The tedium and restrictions allowed for no life, no diversity; she wanted something fresh and living for others.

Qoroxloo was born to forage. To stop would be a death of its own, perhaps in some ways a worse death. She could not surrender and depart; nor would she compete with her family for water.

Three mornings after her last dance, Qoroxloo slipped out. She was followed and joined by a half-dozen women, one carrying a small child, and her son Tsuo Tshao. The tsama melons were gone, and they had to journey far before they came across some roots to dig. It felt good to be out walking, the group later recalled, to be beyond that enforced perimeter, far from the vehicles. Free. Still, as the sun began to set there wasn't much to take back. They decided to camp out that night and gathered wood. Some feared a small fire would be detected, but they lit it for comfort. In the night the baby girl cried out from hunger, and Qoroxloo tried to sing her a lullaby, *xaitsitsoro,* or "I'm feeling cold."

In the morning, the group decided to return to camp. Qoroxloo said she would keep foraging, hoping to find something fresh and moist to bring back for the baby, to ease its crying. Tsuo Tshao, her son, stayed with her. There was no elaborate farewell. The group just went in different directions, Qoroxloo and her son walking into the bush, with her still singing.

The two did not return that day. That night, Bushmen later recalled, there had been much activity as vehicles prowled the perimeter. They heard occasional bursts of gunfire let off into the night, but had no idea whether the antipoaching scouts were hunting game, scaring off lions and hyena, or simply reminding Bushmen of their presence and proximity.

Late on the third morning, Tsuo Tshao staggered into camp, alone and parched. He had been running and appeared afraid, and collapsed in the shade. The group took a bit of emergency water from hiding places in the huts and gave him tiny sips. They made him drink slowly; just enough so that his blood could thin out, cool down, and let him speak. He told them that he and Qoroxloo had been chased by the patrol and had split up. He was not sure what happened to her, where she was, or in what condition.

In desperation, a few Bushmen approached the government patrol. "They thought we were coming to ask to depart," said Mongwegi. When instead they sought to track their matriarch, who refused to leave, the patrol reluctantly agreed to help search, but only with its vehicles, under its supervision, after lunch.

Hours passed. Eventually three officials got into the cab and loaded three Bushmen, carrying a few plastic bottles with water that might revive Qoroxloo. The truck drove in the direction from which Tsuo Tshao came, then the Bushmen leaped down and picked up her trail, leaving their water in the truck bed.

Qoroxloo had been wearing old, cast-off sneakers, the sole peeling away. They recognized the imprint and discussed her movements, reading her speed and pattern or mood just as they had so often tracked a distressed or wounded beast. Her trail moved in an unusual pattern: Steps were spaced close together until they came to the perimeter, where fresh vehicle tracks rolled. Then her path quickly backed away.

November is always one of Botswana's hottest months in the Kalahari's core. Heatstroke can afflict anyone but is especially devastating to females, to the old, and to those actively exerting their bodies under extreme stress.[12] Even a breeze only offers relief up to ninety degrees. It was more than ten degrees hotter as she walked in the desert.

High temperatures were claiming thousands more casualties each year, even in the rich urban world,[13] as an aging and sedentary baby-boom population collided with the extreme heatwaves caused by global warming.[14] The victims experienced swelling extremities; dizziness; hot, red skin; rapid, weak pulse; quick, shallow breathing; exhaustion; nausea; convulsions; and sometimes they fell unconscious. Whereas plunging temperatures shut down the body very slowly through hypothermia, extreme heat brought fast and irreversible consequences. "If you don't stop generating heat—seek shade, lie down, cease your muscular contractions—you will eventually overwhelm your internal thermostat," wrote Peter Stark in his unsettling narrative of all the ways adventurers can die. "Your body's temperature will then go haywire. Your cellular metabolism will accelerate wildly, generating even more heat. A fairly predictable sequence of events follows: unconsciousness, convulsions, and death, as the body essentially cooks its own flesh from within."[15]

The first rule of rescue: Don't put yourself in jeopardy in trying to rescue a victim. The second rule, in the Kalahari: Don't trust or leave your water with men who want you to surrender, and who then drive off without warning, leaving you stranded as the sky begins to darken. Now waterless, the Bushmen search party had to walk back to Metsiamenong on foot, but that night they believed that the patrol must have radioed the news of the missing woman, for they later recalled hearing "flying machines overhead that night" and seeing "headlights driving around." Few people slept much, if at all. It bothered them to know Qoroxloo was somewhere out there, alone with the hyenas and lions, possibly still hanging on.

By morning, a larger patrol appeared—this one more subdued and dedicated to the search. Six Bushmen picked up the trail and continued on foot, keeping water firmly in their grasp. Based on the tracks, they believed Qoroxloo had been alive the day before. But four days in the Kalahari without water left few chances. Her sons hoped they might reach her in time. They could not imagine what she was going through, what she felt.

The time it takes to die from dehydration varies greatly, based on external factors—aridity, stress, heat—and the individual's will to live. Based on those studies of American soldiers in the desert, Qoroxloo striding in one-hundred-degree temperatures needed to drink one gallon of water for every twenty miles. She should have expired after seven miles. But the trackers estimated that she walked without water for several days, covering a distance of more than ten miles.

"Then I saw a figure at the base of a thorn tree," recalled Mongwegi. "We rushed up and found her, lying on her left side, curled up around the trunk in a fetal position."

He lay down around a similar-sized tree to illustrate, rested his head in the sand, relaxed his body, and closed his eyes for a moment. Then he sat up without dusting the sand off himself and stared out at the Kalahari.

"We hoped she was just collapsed or sleeping. But it was too late."

The trackers decided when she had succumbed the previous day.

How did they know?

Qoroxloo had been moving with the shade, and stopped here, when the shade was there, then didn't move as the shade went on, leaving her exposed to the elements. It was a "mooki" tree, with leaves too thin after a long dry season and four years of drought to keep the sun from beating down on her.

Her antelope skin bag held food that Qoroxloo could have eaten to ward off starvation. Evidently she was saving it for others.

ON THE OUTSIDE of the reserve, Botswana's forensic pathologist, Dr. S. A. Mapunda, performed the most critical autopsies on behalf of the state.

On November 11, 2005, the state handed him a politically sensitive case. He was to open and investigate the body of a female Bushman.

The circumstances surrounding this corpse had grown complicated. Dr. Mapunda's autopsy would take place under state supervision, in police presence, at the behest of his boss's boss: President Mogae. It was a delicate situation.

There are a million ways to die in Africa. In rural areas the snakes, hippos, big cats, and crocodiles fatally bite one hundred thousand on the outside while invisible parasites and viruses devour ten times more from within. Cities give rise to suicides and traffic accidents and choking on fried gobs of fast food. In a career that spanned half a dozen countries on three continents, Dr. Mapunda had documented several thousand deaths, and vividly remembered the most interesting cases.

Decades earlier, when a mentor urged him on his career path, Mapunda didn't hesitate. He became very good at what he did, trusted and respected. While 58 million people died each year, far more were born; yet birth varieties paled in comparison with death's infinite options. A certain unique end, with no possibility of return: Mortality made us human. Entering the morgue let Mapunda seek and provide answers to life's most important event, and no office day was dull or routine. "I enjoy my work," he told me with assurance. "I enjoy what I do."

The autopsy chamber still lacked air conditioning and would soon be borderline stifling. The stagnant heat irritated Mapunda, and he had complained on a regular basis. But the hospital administrator was coping with other priorities, like two out of five adults afflicted with the Human Immunodeficiency Virus, the world's highest infection rate.[16] So Mapunda adapted, taking comfort that each corpse rolled up still refrigerated, dissipating the heat, cool to his touch.

Mapunda washed, changed into the disposable robe, and put his medical gloves on. He acknowledged the police who came to attend his examination, standing in the room behind him and across the table. Their routine presence did not make him especially nervous but further added to the heat. He began to sweat into the fresh hospital gown. The hospital could afford no tape recorder or stenographer; without photos or meticulous notes, his observations might blur. So he greeted the attendant morticians assigned to him, grabbed his notepad, and summoned the first corpse.

A gurney rolled in. Atop it the deceased lay enclosed and zipped up in a thick black polystyrene bag. Dr. Mapunda watched his assistants lift and move the relatively light body onto the table before him. Before they unzipped the bag Mapunda grew mildly excited, curious, and scientifically interested. This death held mystery.

All he knew was that the corpse had been found alone in the middle of the Kalahari, a landscape that the government claimed to be uninhabited, sealed off even to tourists. Stories had conflicted about what took place. Activists from Namibia, South Africa, Germany, Switzerland, and France demanded a full investigation, accusing Botswana of murder; the government in defense claimed she died of "natural causes." The local press and foreign media—CNN, the BBC, Australian Broadcasting, Al Jezeera—all swarmed around the case. With international diplomats and media pressing their noses in, tight-lipped officials told Mapunda only that "the woman had been out looking for firewood. She had gotten lost. She died after a few days." That was all.

After several days' delay since death, the body arrived with urgent instructions from the highest levels to examine it. Whenever possible, Dr. Mapunda would visit the scene of each death before examining the body, but in this case it was too remote, too expensive, and too much time had passed. So his conclusions would rest entirely on a detailed autopsy. Mapunda had to be more exacting and meticulous than ever before. Stick to the bare-bone facts as they revealed themselves. As a scientist he would try to tear down emerging opinions and disprove his own assumptions. Only if he failed would the hypothesis stand. "My postmortem examination is not complete," he explained, "until the cause of death is known, the mechanism of death is known, and the manner of death is known."

Dr. Mapunda had authority from the government to investigate, and didn't need legal consent from the next-of-kin. What he needed was confirmation he had the correct corpse. Tsuo Tshao, dressed in tattered and filthy rags, shuffled into the room. His shoes barely remained strapped to his feet. He removed a too-big-for-him baseball cap and adjusted his eyes to the artificial lights, the sterile and antiseptic surroundings, and the metallic glare of exposed and polished steel. It was all so strange to him. He had never been inside a square room quite like this. A mortician opened the black bag.

Tsuo Tshao saw the woman who had brought him into the world, who had sung to him softly, who had taught him the names of plants and which to eat, drink, or use to heal. He gazed for the last time at the wrinkled face that had never yielded, the body that exemplified how to endure when you were isolated for no apparent reason, and recalled how she had been just days earlier, when they were last together, striding across the sands, singing. Looking up, as the morticians moved to usher him out, he nodded in confirmation. "Yes," he said. "It is my mother. It is Qoroxloo."

In American morgues, autopsy rooms often display a sign that reads, in Latin, *HIC LOCUS EST UBI MORS GAUDET SUCCURRERE VITAE*. Translation: This is the place where death rejoices to teach those who live. Dr. Mapunda told me he liked that saying. He smiled and laughed his low, staccato laugh as he translated the words in his head. "Death rejoices to teach the living."

Qoroxloo might have objected to being sliced and explored, cut to the quick in the name of science. Her people did not practice human dissection.[17] Yet over her lifetime she cut open hundreds of antelope and discovered in those de facto autopsies their inner physiological workings; she could draw comparisons to her own people, and while alive, Qoroxloo had passed on knowledge that helped. Now, perhaps, she could do so in death. If autopsy brought her family answers, peace, strength, truth, Qoroxloo might well have rejoiced.

Inexperienced pathologists might rush to cut, but Dr. Mapunda delayed touching a scalpel. The mood was calm. The procedure would be carried out with respect and seriousness where the "law is at stake, and in which foul play may be expected," he said, adding that there was no such thing as an inconsequential detail. Mapunda followed a strict three-stage sequence, starting with meticulous visual observation from head to toe, missing nothing. He pulled the zipper slowly and steadily, holding it away from the body, making sure not to let it snag on clothing or skin. He described the attire, inspecting for abnormal human stains: blood, urine, feces, semen, grease, burns—anything that might shed light on what took place at the time of death. He looked for burst buttons or seams that might suggest force or violence. Qoroxloo's clothing was smudged, smoky, torn here and there, but otherwise intact. There appeared to have been no struggle.

The dead tell no tales, but their belongings just might. A key chain, crucifix, or lapel pin could unlock the deceased's habits, lifestyle, beliefs, and activities. A purse could reveal friends, family, recent contacts. But after a quick yet thorough search, no valuables could be traced from the attire, Mapunda observed. Qoroxloo lived without closed doors, locks, phones, wallets, coins, or currency. She could not count, and where she lived there was nothing that money could purchase. "She had no pockets." This in itself was subversive. If you don't seek to acquire, you don't need money. If you don't need money, you can't be voluntarily persuaded to do what those who have money want most from you.

Dr. Mapunda scribbled down the place, time, sex, and age of the body, but even simple questions proved tricky. Born "beyond the beyond," Qoroxloo kept no calendars, birthdays, or historical records. Botswana's census tracked down citizens, took snapshots, and assigned them a number next to a name and age. She had no idea how old she was. The census taker had looked at her wrinkles and estimated her birth to be 1924. Mapunda now wrote "female, aged 81."

Next the autopsy form asked for "race." "Nonracial" Botswana defined Qoroxloo's band by negatives: what they lacked, where they were not, how they lived wrongly, their abnormal language or skin color, or when they arrived. Yet there was no official space for "Bushmen," "Basarwa," "San," "Remote Area Dweller," "Hottentot," "Noakwhe," "First People." Even in death she could not be assimilated.

Mapunda cut away an old turquoise and beige scarf covering her head. Nappy and brittle to the touch, her hair broke off before growing too long. It was distinct to Bushmen and had an evolutionary role: First it protected the skull and its precious contents from the convection heat of the sun; then it grew stiff to allow the skin to breathe and cool. Scientists believed this kind of hair, so perfectly adapted to Africa, was common to all humans before our ancestors migrated into cooler, cloudier regions of the planet. The cheeks had grown hollow. Dr. Mapunda "made a note of the blood stained oozings from the nose due to early decomposition."

The body was clad in an old, tattered and torn light blue dress and grayish shirt. He removed knotted, ropelike fabrics slung across her chest, and "the wrists carried laces of some sort": six white ones on her left, three black ones and a fourth white on her right. Beneath dry, shrunken, wrinkled skin, her muscles were taut and healthy from a life of almost constant motion and walking upright over great distances. Vertical posture dissipated heat, reduced solar exposure, and accessed more favorable wind speeds and temperatures, thus conserving energy. As climate grew drier, habitats converted from forest to grassland savanna; surviving hominids had to travel longer distances to gather food with carried tools, as bipeds. But time also sculpted these legs to run, to chase down and exhaust prey to death through dehydration, dry starvation, and shock.[18]

Dr. Mapunda examined her skin for cyanosis, which indicated suffocation. He probed her eyes to see if they were pale, jaundiced, bleeding into the sclera. He investigated ears, nose, mouth, chin, neck in the same meticulous sequence. Upper limbs: left then right. He checked her chest for any abnormalities: breastbone, ribs, lining of ribs, breasts for bruises or injuries. He pored over her abdomen, her perineum, groin, and thighs. He inspected her external sexual organs, followed by the lower limbs, left then right. His assistants turned the body, and he surveyed her prone then repositioned back supine. At this stage, after documenting any abnormalities, Mapunda could "tell quickly if there was foul play." No outward signs suggested any.

The final stage of autopsy involved internal examination. "You must open the three body cavities, irrespective of the manner, mechanism or nature of the case." It was time to pick up the knife.

Mapunda made an incision of the brain, top down, using a long, pointed, and single-edged brain knife, and took out different structures: first her cerebrum, or big lobes; then her cerebellum, or small lobes. "You see a desiccation of features," said Mapunda. "A drying out. The moment you remove the membranes, you usually see an oozing out of cerebral spine fluid. But in this case it was all dried out, shrunken."

That condition marked her entire body, inside and out. Qoroxloo's dry skin showed "loss of elasticity or turgor." It was "adherent to bones because of less subcutaneous cushion, or fat." Her nail beds projected excessively as fingertips and toes shrunk, giving a false sense that the nails were longer. Dry lips. Dry eyes. Dry connective tissues. Her mouth revealed a dry tongue. In clinical cases scientists might wash the body for a clean look. "But we don't do that," he said. "All this time we've never opened a tap of water. Every bit of dust that falls off is important."

As Mapunda moved to the abdominal cavity he tested an emerging hypothesis. Normally intestinal contents would be watery; heat turned fluids gassy, bulging outward, convex, bloated, and straining against belts and buttons. But Qoroxloo's midsection was, abnormally, concave. It suggested there was no fluid inside whatsoever.

Cutting her open confirmed this. When he looked for signs of dehydration, he found them everywhere. The liver stored little, unless someone suffered for a long time; when depleted, as in this case, "it gets replaced by fat cells." She had a shrunken spleen and shrunken pancreas, although otherwise healthy. The usually thick intestinal walls were in Qoroxloo's case "thin-walled and transparent," without contents and minimal gasses. "The stomach's interior and posterior walls were touching each other," empty. Her kidneys were pale, with no contrast between the outer cortex and inner medulla; the abnormality indicated they had been "shocked."

To better assess Qoroxloo's shock, he opened her chest cavity. It was here, in what some considered the seat of emotions, that shock was most pronounced, for it marked a circulatory failure. He made incisions to the lungs, and from cut surfaces documented the minimal edema he found. "Much more air than fluid. Minimal secretions in the upper and lower air passages. They are dried up."

The water content in her lungs and blood vessels was "concentrated, and showing a high viscosity." Her blood had slowed, dragged. Fluids could not have flowed at the rate her body demanded. Because her thirsty tissues could not be replenished, her inter- and intra-cellular balance of electrolytes was disrupted. Her body went into circulatory failure. "This is what kills," Mapunda explained. "The body has built-in adaptive responses. But shock takes place only in the compensatory stages of survival. As the provoker continues to exert the negative impact, and the body goes into de-compensation, the worst happens."

He depicted "shock kidney, shock lung." Then, after making coronary incisions, he opened and described, in graphic detail, what he saw in Qoroxloo's heart.

The human body is an extraordinarily complex organism. It is often portrayed as existing only in competition with other humans. But as water grows scarce, the body competes even with itself.

Cells vie with each other over the neutral water they all require. Voluntary muscle tissue contends with involuntary tissue for water. Blood consists almost entirely of water. The heart, stripped of all romance and mysticism, is essentially a pump. During her life, Qoroxloo's pump beat more than 2 billion times; its entire role was to replenish the body with enriched water to satisfy the thirst of organs, including the heart itself. Each organ claimed supremacy of need, of course, especially tissues of the nervous system. (More than 90 percent of the active human brain consists of water.) That meant Qoroxloo's heart and brain were locked together as thirsty rivals.

Mapunda opened the opaque sac that protected Qoroxloo's heart from friction, severed smaller blood vessels long known as heartstrings, lifted her tough rubbery muscle the size of a fist, cut into her right atrium, and sliced through the sinoatrial node—the cluster of cells which set the pace of Qoroxloo's pulse.[19] What he saw inside her heart was unusual and disquieting.

Mapunda had come across a few other dehydrated corpses in his career: children wracked by convulsive diarrhea, unable to retain water; adults who had died within days. But Qoroxloo's death had not come early, fast, or sudden.

Qoroxloo showed "a brown atrophy of the heart: small in size, dark brownish in color." And within her heart chambers, the muscle revealed "the pigment accumulation of lippochrome." His characterization, "brown-heart," only occurred "through chronic or longstanding suffering of dry starvation and dehydration," he explained. "You need time for that type of arrangement. It doesn't happen quickly."

This indicated that Qoroxloo's heart had sacrificed itself in favor of sending more water to her brain. It also meant the lapse of time of suffering for those changes to happen took more than three days. It dragged out over many weeks, months. Years.

His autopsy complete, Dr. Mapunda left his morticians to clean up Qoroxloo's body. Finally, they could turn on the tap. Qoroxloo's tissues and organs were irrigated, absorbing moisture like a sponge. In death the government bathed her in the abundant waters denied through the very end of her life.

On "4 November 2005," Mapunda reported, "Death was due to: 1a. DEHYDRATION AND SHOCK; and 1b. DRY STARVATION." In the former, Qoroxloo's body was rendered dry due to lack of or insufficient intake of fluids, causing circulatory failure; the latter when her body was denied food and fluids for many days. Dr. Mapunda professionally satisfied the tasks required of an autopsy. He provided "the aetiologically specific *cause* of death" (dehydration); the "non-specific common mechanism of death" (shock). His third answer, concerning "the manner of death," remained problematic. Qoroxloo showed no signs of disease; aside from water deprivation, her tissues appeared healthy, strong enough for years of life. So Mapunda failed to determine whether her "unnatural" death was accidental, homicidal, or suicidal.

Once again Qoroxloo's death, like her life, refused to fit into any neat, clinical category. Some observers have written how elderly Bushmen accept that in times of great thirst their band may need to leave them behind with no more than a fire to warm them and a circle of dry thorn bush to ward off hyenas until the end comes. They do this to avoid dragging down others to share their fate.[20] That may be true. But if so, in this case it was Qoroxloo who left the band behind with a fire to protect them, while she reduced pressure and risks on them.

Dr. Mapunda knew Qoroxloo's political ordeal in the heart of the Kalahari no better than Bushmen knew the physiological struggle that raged silently within her heart. Only after his autopsy was shared with the Bushmen could external and internal evidence coalesce into truth. His report showed Qoroxloo's dehydration took months and years. During those years, as the government denied water to Bushmen, her band watched Qoroxloo deny herself so that others could eat and drink—the ultimate form of human adaptation. She never mentioned her austerity or abstinence. She never complained. Qoroxloo's self-sacrifice came to light only after her death, when her chest was cut open and revealed dark brown layers of her inner coronary wall, a condition best described as a dry heart.

CHAPTER 22

Release

WHEN QOROXLOO'S DRY HEART STOPPED beating, it was unclear what became of her spirit. Faiths vary among Kalahari bands, and no two Bushmen share exactly the same beliefs. Even Qoroxloo's band had been exposed to missionary influence until lines blurred between traditions; Botswana's Zionist Christian Church had converted her grandchildren hundreds of miles away, where there was enough water for baptisms.

But in the reserve, the old beliefs endured and were honored. The journey of the dead resembled less that in the Bible than the shades described by Homer. When Kalahari Bushmen died, their souls were not automatically assumed to ascend up to heaven; the ancestors carried on down beneath the ground, lingering close to where they were ceremoniously buried.

A Bushman once described that subterranean abode of spirits. "From this sand downward," he explained, it was "soft, like the sand itself is." Then came a hard layer of stones—like calcrete and diatomite layers in pans. And then at last came "water; water, water on, downward, endlessly."[1]

That aquatic underworld fused the real and the metaphysical. But visions of a deep underground lake (Egyptian Book of the Dead), well (I Ching), or stream (River Styx) in the afterworld imprinted on more than just these parched Kalahari souls. It was rather a *primordial image,* or mental pattern so deeply embedded and shared by so many cultures, that Carl G. Jung considered it a universal archetype that animated the psyche of all humans. The mythological "deep body of water" even encompassed Sigmund Freud's negative repository of repressed personal emotions and desires in a more powerful "collective unconscious." Perhaps a life of constant struggle to replenish our bodies drove humanity to shape a watery afterlife that forever quenched the thirst.

Yet this same afterworld might also use water to torment. The word *tantalize*—to tempt with a desired object kept just out of reach—came from the Greek figure Tantalus, condemned in Hades to stand in a pool of water that receded whenever he bent to drink.

Bushmen tales have linked water to the afterlife, perhaps dating back to prehistory. Stories about the spirit world were passed down across generations, described around the fire, or inscribed graphically through rock wall paintings or engravings. These images have endured longest in the driest, isolated places. They have been widely discovered—in sealed European caves like Lascaux and Altamira; in aboriginal Australia; and in the cultural remains of the Anasazi in the American Southwest—and while Bushmen rock art showed large mammals amid scattered human figures in what was long assumed to convey the story of a hunt, more complex interpretations recently have emerged. Often paintings were strategically superimposed or showed bleeding noses; more than a third involved grids or lattices, parallel lines, dot clusters, zigzags, concentric circles, or U-shaped meandering lines. What did it all mean? Since no Kalahari Bushmen painters still survived, researchers began to assume the meaning behind rock art belonged to another mind-set and had been lost forever. Then a bold interpretation broke through the wall to interpret its inner side. These pictures did not reflect everyday life, argued David Lewis-Williams, or even merely religious expressions; they portrayed transrational visions of the Bushmen otherworld: a painted record of their visions, insights, and spiritual struggles, many of which had to do with rain. Through exhaustion and hyperventilation in a ritual dance, Bushmen like Qoroxloo entered in an altered state. In that trance, many would describe a feeling of diving into water holes or traveling underwater.[2] And there they fought for the lives of the sick, killed rain animals who lived in water holes to release the clouds,[3] and began a journey through a gateway into the transcendent realm of subterranean water.[4]

In this they were not alone. Indigenous people ranging from Canada to the Amazon and Australia spoke of diving into holes in the ground, swimming in the bowels of the earth, to rivers flowing in opposite directions, to access this subaqueous spirit world.[5] Perhaps this archetypal, mythological subterranean water—which remained so maddeningly out of reach throughout the hot days of the living—represented to the Bushmen an otherworldly salvation, a form of unearthly paradise.[6] Even so, for access to this afterworld Kalahari Bushmen required a portal for the dead, which in turn demanded a prompt and proper burial at home.

FOR WEEKS, THAT funeral appeared unlikely. Outside the reserve, a bitter and intensely politicized struggle was shaping up over possession of Qoroxloo's corpse. It resembled the international controversies that took place a decade earlier, first when South Africa fought France to repatriate and bury with dignity the preserved remains of the iconic Bushman Sarah Baartman, known as the Hottentot Venus,[7] then when Botswana reclaimed possession of an embalmed Bushman who had been on display in a case in Spain and gave the body a state funeral.[8] Only now, ironically, it was Botswana that wouldn't surrender a Bushman's remains.

The tug-of-war began within hours of finding her dead, then escalated after the autopsy, and was mainly fought over the question of whether mourners could have access to freshwater.[9] Officials demanded a list of exactly who the family would choose to attend the funeral. Bushmen refused to produce one, saying any friends or relatives were free to grieve. Next the government set conditions over how people would travel in and honor their dead, and when. Again, Bushmen refused to comply with its twenty-four-hour timetable. In a prepared statement, Botswana tried to limit the water that mourners could bring into the reserve, restricting the quantities to only an amount "adequate for personal use" and warning: "Any excessive amounts of water and foodstuff brought in will be confiscated."[10]

As diamonds' sparkle was tarnished by the ghoulish scenario, De Beers' new chairman tried to distance his brand from the obstructed funeral, urging Botswana to end the Bushmen evictions. Mogae refused. "To be honest [De Beers] did not bring anything new to the table," scoffed the president's spokesman Jeff Ramsay. Pushing back, he added that the only acceptable outcome was for Bushmen to drop their lawsuit, call off the protests, and learn to bury their dead outside the reserve.[11]

That was something Qoroxloo's kin refused to do. If exiled in death, her spirit would linger too far from home, taking healing strength with it, eroding the powers beneath the Kalahari, and surrendering a birthright. Like all of us, Bushmen differed over the nature of the afterlife. But they believed that something would go out of them as a people if their ancestors were dispersed too far from home.

Then an interesting thing happened.

In a country wracked by AIDS, where families might attend one or two funerals each week, people could easily imagine themselves in a similar situation, and soon the government's ploy to blame Bushmen for the burial delay began to backfire at home. Six weeks after Qoroxloo's death, one Botswana citizen Mpho Sekute urged in a public letter for President Festus Mogae to treat Bushmen as he treated everyone else. Why, she asked, should Botswana "delay the burial of Mrs. Duxee and deny members of her community their right to bury and mourn her? Where and when in Botswana has the government ever dictated who should mourn a deceased person and how? What will the government lose if it allowed Boora [*sic*] Duxee and their friends to burry [*sic*] and mourn their mother?"[12] Enough was enough. Let the Bushmen put Qoroxloo in the sand, return her home, and release their ancestor into the afterworld.

Whether from internal pressure from De Beers or external shame from voters, Mogae finally relented. Not only did his General Merafhe allow an unlimited number of Bushmen to join Qoroxloo's funeral procession; he even provided transport into the besieged land.[13] Some sixty adults, with dozens of children, were trucked in from the relocation camps of New Xade, Kaudwane, and Kedia. The scales of power had tipped. When the government wanted to film the funeral as evidence of a dangerous assembly, Bushmen stood firm until sheepish officials put away their cameras. Women were reluctantly allowed to gather the right plants to be burned. And while government officials vowed to arrest any men who hunted antelope, they looked the other way when Qoroxloo's grandsons brought back a steenbok for the funeral. The gathering was a bittersweet reunion. Some Bushmen had not seen close family members for several years, and that night they sang and clapped by the fire in honor of Qoroxloo.

"She raised us and took good care of us all the time, and she never came out of the reserve," said Bilathwe, recalling her grandmother. "She was a quiet woman. After she lost her life, still she is left through the ancestors. She's not gone. She's still around to be asked help and guidance from the living."

"Qoroxloo was the person who made me who I am," said Motswkgakala, her nephew. "She was a very kind person to everyone."

Kathameseng, her cousin, agreed, but felt her loving was almost too strong. "She relied too much on herself. She taught very different kinds of dancing, and telling jokes. A brave woman, she used to hunt with my father."

Setheke thought of her great-aunt's determination. "She always walked hard even though she was grown up and old. She raised her kids in the dry times when there was no water whatsoever, but she managed to keep them and raise them. All that was because of her skills, and, of course, the help from ancestors."

Galomphete, her grandson, remembered Qoroxloo as especially "welcoming, with a sense of sharing. She loved to be with people. She was very good at making jokes, so I loved staying with her. Sometimes she scared you, then after she would laugh at you and that made her very special."

Belesa loved his wife, he said, "because of the way she behaved toward people; she did not discriminate against anyone." But he regretted never having had the chance to say good-bye, tell her he would miss her. Belesa wanted Qoroxloo to be remembered by things she had been doing before she died, things like gathering, hunting, and dancing. "She was the strongest in many ways," he told me. "She could even work in sunny conditions, and she took good care of me and the kids. I did not ill-treat her; that made her so excited. Her dreams were seeing her children becoming like her, being strong and brave. She also dreamed of seeing and living in a free Kalahari reserve."

The Bushmen did not have time to prepare everything they felt they needed. Against tradition, no one could look in and touch the body of Qoroxloo, which had decayed too far over the long hot weeks since her death. Nevertheless, they put her in the ground and scattered what they could over her, and covered her mound with thorn branches, and blew smoke and hoped she would understand how hard they had tried to honor her under the circumstances imposed on them. Qoroxloo's funeral took place on December 25.

The following day the sky exhaled.

Something somewhere was released as the four-year drought broke. Clouds cracked open in violent thunderstorms as the rain fell down in big drops, dampening the thorny branches covering her grave, dripping down, forming small puddles, penetrating the membrane of the surface world, before the rains sank down through her soft sand grave to join the subterranean abode of *g/amadzi* filled with the water; water, water on, downward, endlessly.

PART VIII

Resolution

CHAPTER 23

The Verdict

THE RAINS FELL AND FELL ACROSS THE COUNTRY, and in the city people inhaled the smell of steam rising off the hot asphalt as sheets of water ran off into the gutters.

In the Kalahari, the clouds emptied themselves, replenishing millions of dispersed natural pocket reservoirs. Wild water filled the shallow pans; the sip-wells; the hollows of trees; the cracks in the rocks. Sinking drops were seized and absorbed into the roots and then burst forth into the flesh of fruits, melons, tubers. Rain restored green to the native thorn bushes, trees and grasses sprang to life within days, the wet energy and sap flowing into the rodents and antelope, whose blood in turn fed the coiled muscles of snake, cheetah, lion, and hyena. Omnivorous Bushmen watched the rains fall on dry sand and again felt secure.[1]

As they fell, the long reluctant rains cooled more than Botswana's landscape. They seemed, for a while, to soothe tempers and reduce heated outbursts. Government officials breathed out relief as the slack waters behind Gaborone Dam began to inch up against the impoundment, slowly rising to overflowing once again, and thus restoring the source of state authority, wealth, and the country's currency, *pula*.[2] As shared rivers flowed, tensions eased among Botswana's rivals. Neighboring states became bighearted once again and forged tighter agreements over the Okavango, the Limpopo, and the Zambezi.

Internal restrictions did not end for another six months,[3] while the government kept domestic controls tight.[4] And long after the drought broke temporarily, the Kalahari siege remained in place permanently.

Officials retreated to the periphery of the reserve and reduced their harassment of Metsiamenong to irregular visits every three weeks or so. But they continued to accuse the Bushmen of making bombs, and wildlife scouts routinely arrested and beat up hunters. There was no truce, but no further casualties either as the Kalahari microcosm seemed to be waiting for a final verdict to decide its fate.

When the rainy season ended, the high-court case resumed. With the end in sight, the government called its final witnesses, and when that was done both sides made their closing arguments.

The Bushmen argued that their rights to reside freely in their reserve were first established and recognized in 1885 under the British Protectorate; those exclusive rights remained following Botswana's independence and were reinforced by the constitution. By cutting off water, Bushmen claimed, the government abused its wildlife regulatory powers to force them from the reserve. The government broke off negotiations and unilaterally drafted its own plan that airbrushed the Bushmen out of the reserve. It pretended that those who remained in or returned to the reserve simply did not exist, and tried to make it impossible for residents to remain. No evidence supported government rationales for its conduct. Finally, whose development was at stake, Botswana's or their own? "The test of a mature democracy," concluded the Bushmen attorney, Gordon Bennett, "is its ability to tolerate and respect the choices made by its minorities, and to resist the temptation to impose upon them a way of life they may not want and do not seek."[5]

President Mogae's government defended its actions as emphatically benign. The primary concern was to protect viable wildlife populations from being decimated and infected; human residence within the reserve posed a clear disturbance to flora and fauna and ran counter to the policy of ecosystem preservation. Also, the government desperately needed to reduce the prohibitive cost of water provision to an acceptable level. Water services in the reserve were too expensive to maintain on a long-term basis, explained Sidney Pilane, so condensing people into a less remote and scattered location was by far the best way to economize. Above all, he added, Botswana constantly sought to introduce the Bushmen into the mainstream of modern society; had consulted with them for years about doing so; and had given them six months' warning about the end of water. Even then, government officials never terminated water supplies; they had merely relocated them elsewhere, outside the reserve. Finally, Pilane added that Botswana could not have cut off water as an ulterior motive of force, since Bushmen like Qoroxloo chose to stay, and dozens went back despite the lack of government water.

On September 8, 2006, in elaborate arguments filling 750 pages, both sides rested their case. At that point, the high court's three justices bemoaned how what had been initially an urgent application filed in haste on behalf of 240 citizens on February 19, 2002, had become over nearly five years the most expensive and longest-running trial of a scale that none of the parties, nor the two courts, could have initially anticipated, attracting more national and international interest than any previous case.

The hype and delay escalated even further. After the high court reviewed the evidence—4,500 pages of legal documents and 19,000 pages of transcripts—its ruling would be broadcast. Live. Globally. Three long months later.

On December 13, 2006, the courtroom filled up early, and usually empty wooden benches groaned under the weight of bodies crammed hip to hip. Dozens had to stand. The crowd showcased Botswana's best side: young and old, black and white, journalists, lawyers, activists, and foreigners all respectfully taking seats beside one another. The collective mood was one of anticipation and excitement.

Alice Mogwe sat with her back support, prepared for a long day. The human rights activist remained saddened that the water conflict had grown so bitter and divisive, gone on for so long, and caused so much irreversible damage to both sides. A diplomat at heart, she wished it could have been settled quietly over tea. And yet, she acknowledged, "until this case, Botswana citizens had never really experienced a sense of history, of shared heritage." I found this a remarkable insight, and true. Neighboring African countries endured cold-war struggles, ideological civil wars, the legacy of slavery, colonial subjugation, racial oppression, and class struggle. Botswana mercifully had escaped all that, sailing breezily from benign protectorate to independent middle-class nation of strip malls and suburbs. The only forces that made the country stand out as truly exceptional were its aridity, its wildlife, its Bushmen and its diamonds. All had come together, clashing over water. "Now, perhaps," she concluded, "we will have something we shared that better defines us, and shapes our sense of history."

Behind the attorneys the courtroom seats were packed with rivals who had fought outside the courtroom. When it seemed the room could not get more crowded, the Bushmen plaintiffs filed in, one by one.

Some hitched rides down from New Xade and from Kaudwane. All had camped out in the fields around the courthouse, taking food and shelter wherever they could. Dust covered the animal hides that some wore over hand-me-down clothing. The smell of their stale sweat and woodsmoke joined air that had been filled with perfume and aftershave. They huddled close to one another, eyes wide in the camera flashes, but remained quiet and stayed patiently in line. Finally, someone found folding chairs, and Bushmen filled two rows, waiting to hear their fate.

Three judges ushered in, faces inscrutable, and took their chairs. Then Chief Justice Maruping Dibotelo began to speak.

AS THE ELDEST and most conservative of the three, Dibotelo had insisted on taking frequent tea breaks, and his relationship with President Mogae paralleled that of Justice Antonin Scalia with Vice President Dick Cheney. In this case Dibotelo looked into the record and found abundant evidence to support five of the government's seven positions. If an official said water was too expensive, no further facts were necessary. Since several told the Bushmen that if they didn't move, the government would terminate water, that in itself seemed ample consultation. If some Bushmen who did move drank from government taps outside the reserve, water could not have been used as force. What's more, since Bushmen like Qoroxloo remained inside the reserve, termination of services couldn't have been coercive. Bushmen possessed their homeland inside the reserve, but they weren't forcibly or wrongly deprived of it. While it may have been unlawful to prohibit Bushmen from hunting, he didn't think it unconstitutional. Indeed the case should be dismissed, he said, on grounds that the state owned the Kalahari, and all water beneath: "The contention of the applicants that the government unlawfully deprived them of their land, must fail." The elder judge suspected the five-year litigation would not end asymmetric water war, but "hoped that, whatever the outcome of this case, the parties will . . . come together to resolve their differences."

When his verdict was translated, the Bushmen crumpled. This was a blow to the hopes of the one thousand wanting to return home. Roy Sesana called the opinion unfair and cast doubt on the court's impartiality. "We are very sad about this statement, they [judges] were told how to give the verdict, they were blinded and had their ears covered."

The BBC rushed outside to be the first to break the news, live: Bushmen lost!

That story turned out to be premature. Dibotelo had only offered the first of three judges' opinions—all of which would be tallied up into a final verdict—and the next ruling came from Unity Dow.

If you crossed Justice Ruth Bader Ginsberg with John Grisham and Africa, you might get someone resembling Unity Dow. Born in a rural part of the country, Dow was educated abroad and returned as a women's rights activist, the country's first female high-court judge, and a wife and mother who in her spare time wrote legal thrillers that explored poverty, child abuse, rape, murder, and the struggle between African and Western values. She seemed so inclined to sympathize with Bushmen that, as she began speaking, the government observers stopped taking notes. But she caught them off guard, first lashing a Bushmen attorney for grandstanding, then rebuking Sesana for refusing to testify while complaining "through the media and without the limitations of an oath to tell the truth." Then, in substantial detail, Dow described Bushmen blood and marriage ties, their lack of literacy, their dwellings, and their perceived dependence on water, noting how they "resembled a group one might see at a refugee camp, bare-footed, poorly clad for the weather, and the desert temperatures do, during winter nights, plummet to freezing, and obviously without sufficient water for proper hygiene."[6] She decried the marginal economic status of Bushmen—called "little people" and "uncivilized" and "wild"—who relocated following the dismantling of huts and emptying of water containers by armed officers.[7] Dow understood how Bushmen saw land and water as "not merely a possession and means of production," or "a commodity that can be acquired" and compensated, but rather in a special spiritual relationship. What's more, Bushmen had inhabited the Kalahari since long before Botswana developed laws. "The policy came after the people." The only purpose of destroying tanks, pouring out water, and sealing the borehole was to force Bushmen out.

> Water is a precious resource anywhere and a particularly scarce one in the CKGR and it would have been brought there at some costs, so to up-turn tanks would have been a dramatic and clear

> statement to the Applicants. This is particularly so since those in charge of the relocation exercise needed water too, but this problem was solved by bringing water that they could control.

Based on this exhaustive foundation of evidence, Dow decided the government knew that water availability determined mobility and use of the Kalahari, so its two-pronged policy—"encourage but not force" and "no water provision"—revealed inherently irreconcilable contradictions. Cutting off water and banning hunting, she argued, infringed on the Bushmen's constitutional right to life, liberty, and freedom of movement. Thus Dow ruled for the Bushmen on all counts: They were dispossessed "forcibly, wrongly and without their consent" by a government that terminated essential water in ways that were "unlawful and unconstitutional." She too urged both sides to reconcile differences. To resolve a dispute, which at first blush is about termination of water, everyone must address a deeper context, the nub and heart of all progress. She concluded that water does not belong to the governments to do what they wish with it. Water belongs to the people.

Reuters dashed off a wire story: Justices appeared divided, deadlocked. The third judge would cast the swing vote and tip the scales of justice.

The Honorable Mpaphi Phumaphi was a wild card, neither as conservatively deferential to President Mogae as Dibotelo nor as sympathetic to Qoroxloo's band as Dow. As a male, he subscribed to conservative traditions, yet did not belong to the ruling Tswana tribe. He was a Kalanga, one of the country's minorities, and thus brought an outsider's perspective on the currents driving Botswana. As he delivered his verdict with a deep, low voice, he clearly relished not only his international platform but his having the decisive last word.

Phumaphi believed the president, not the court, should decide how, where, and to whom government should allocate the water it owned and managed. Based on his philosophical principles, he ruled that termination of water was lawful and constitutional, and the government did not need to restore its provision to anyone. And yet Phumaphi then proceeded to surprise the courtroom. He was genuinely impressed by Bushmen determination and grit, and ruled that Bushmen were in lawful possession of their land in the Kalahari reserve. Despite the knowledge that water would be cut off, they decided resolutely to remain in the reserve long after services were terminated, and reverted to their old ways of survival. Phumaphi saw history repeating itself, as Bushmen—formerly displaced from fertile land with readily accessible water by British settlers—were once again displaced from the last place they could and always had lived. Phumaphi felt something must have suddenly changed Bushmen minds, and that something was water. People just didn't fight a situation they had agreed to voluntarily, he said, so of course the emptying of water tanks was to pressure them to relocate. They were, Phumaphi ruled, thus dispossessed wrongly and without consent. Finally, he thundered that the simultaneous cutting off of water, the stoppage of food rations, and the prosecution against hunting were "tantamount to condemning the remaining residents of the Central Kalahari Game Reserve to death by starvation." These actions violated their right to life, and their right to free movement, and thus his verdict was in favor of the Bushmen.

Combined with scores from the other two justices, in five of seven counts, Bushmen emerged victorious, with the notable exception of a right to water. It was ruled on the thirteenth day of December, 2006.

The courtroom erupted.

Within seconds Roy Sesana strode into the sunlight. He lit a Courtleigh cigarette, performed a jig, smiled broadly, and proclaimed: "We have been crying for so long, but today we are crying with happiness. Finally, we have been set free. The evictions have been very, very painful for my people. I hope that now we can go home to our land."[8]

Survival International's director had long expected Bushmen to lose, but now Stephen Corry hailed what he called "a victory for indigenous peoples everywhere in Africa. It is also a victory for Botswana. If the government quickly enacts the court ruling, then the campaign [to return the Bushmen to their land] will end and the country really will have something to be proud of."[9]

The previously restrained Bushmen attorney, Gordon Bennett, was jubilant, telling the surrounding reporters how the case was a precedent for indigenous groups around the world, from North America to Australia, Asia, and Africa, and that more specifically the high-court ruling simply upheld "the rights of the [Bushmen] to live inside the Reserve as long as they want—and that's a marvelous victory."

Sidney Pilane appeared visibly shaken. For several years he had staked in the ruling not just his own reputation but the president's stature and the country's standing in the eyes of the world. He tried to spin the "mixed decision" as "not at all clear what it means in terms of implementation." But it had been undeniably a scathing judgment against him and a crushing defeat for the government, which found no basis for appeal.[10]

De Beers continued to distance itself from Botswana in order to extract a corporate victory from the verdict. Executives noted the court's unanimous agreement that diamonds played no role in the dispute between the First People of the Kalahari and the government, quoting Justice Dow's dismissal of such claims as irrelevant and ridiculous.

> While diamond mining as a reason for the CKGR relocations might be an emotive rallying point, evoking as it does images of big, greedy multinationals snatching land from, and thus trampling the rights of small indigenous minorities, the case before this court does not fit that bill. It would be completely dishonest of anyone to pretend that that is the case before this court. Those looking for such a case will have to look somewhere else.

The next day I went somewhere else to look up a highly educated management official who had asked not to be identified by name and who even then had refused to speak until after the high court's verdict.[11] Let's make this person a woman and call her Hesitant. Four years earlier, Hesitant had watched the Kalahari water terminations and its violent aftermath with growing dismay and a feeling of remorse. At that time a high-placed government employee under Local Government Minister Margaret Nasha, Hesitant knew she had done the right thing for the nation, but regretted the consequences. "It was painful to picture Bushmen being kicked out," she once lamented to a friend, "all because of the mining that will take place." When teased for naively falling for the antidiamond propaganda spewed out by FPK and Survival International, Hesitant paused, midstride in the hot parking lot, and leveled her gaze. "I was there," she said. "I saw the plans." Then she fell silent. By the end of the legal case she was no longer working for Nasha and rejoiced at the high-court verdict until she heard that President Mogae planned formally to accept, yet quietly ignore, the ruling.[12] Since technically the government had not wrongly terminated water, he would refuse to restore it. "I saw on the news yesterday that [Bushmen] would still be forced to bring their own water," she said in frustration. "You can't just pick up and put them among us to struggle in the same place. They know that the dove makes its song here, at this end, this season and time of day. When that's gone it is unhealthy, disorienting." So what was going on? She explained how, after four decades of growth and prosperity, Botswana's citizens were on average getting poorer. Half the country lived in poverty, and the sources of raw wealth were diminishing. "There were two things that we believed cannot be touched," she said. "Our diamonds and our wildlife." Central Kalahari Bushmen were seen as a threat to both.

Only two things left lingering doubt about the thirsty diamond mine rationale. First, Botswana legally owned all resources beneath the sands; so where was the economic conflict? Second, kimberlite pipes, a diamond-bearing geological formation, were the approximate width of a barn, so why did the government need so much space? Hesitant erased both doubts, explaining how resource ownership could always be disputed, which was indeed already happening in other countries. Also, the documents she had processed showed that mining in the reserve, as it became economic, would require lots of space to make room for the entire extractive and processing effort, space for mining and roads, a support infrastructure, employee housing, stores, labor and security and fencing all large enough to cover an area the size of Luxembourg. As the industry was evolving, this was to be an advanced technology mine, processing as much as extracting. "I have seen the plans; I have looked at the blueprints," said Hesitant. "Don't quote me by name, please, as I would get in trouble. But of course the water cutoff has to do with diamonds. It has everything to do with diamonds."

CHAPTER 24

The End of the Beginning

WITHIN WEEKS OF THE COURT RULING, forty evicted Bushmen made it back into the Kalahari, ignoring President Mogae's attempts to make them stay outside. By April, two hundred had returned. Despite dual pressures mounting against them—an increasingly hot sun and vehemently hostile officials—they reinhabited the core of a waterless reserve at the center of a landlocked country within an increasingly arid subcontinent. For every Bushman caught, accused, arrested, and roughed up, several others sneaked in to gather or hunt, preferring to live freely without official help, without water that had strings attached.

Two of Qoroxloo's grandsons were the first ones back home inside, where they reside to this day. "As usual the government is doing everything it can to see to it that we resist from going home," Galomphete recently wrote me, in a note about life in the Kalahari that understandably took some time to reach the outside world, "but that's not a surprise. We will keep fighting."

In the note he described a recent hunt with his brothers. The three young poachers were joined by Mongwegi, Kalakala, and Tshokodiso—the men who had tracked down and recovered Qoroxloo's body and who had dug her grave—and by Mohame Belesa, her widower. "As it is our custom," he wrote, "we went hunting thinking of the people in Metsiamenong because we felt we had to do something for them."

The hunters chased down and killed two gemsboks and distributed the meat. Bushmen ate the flesh from one of the Kalahari's most desert-adapted antelope, a beast that survives by digging up the moisture embedded in roots and tubers and leaves. The feast and reunion and dance in the night helped close the circle once more, reconnecting those returning home with those who never left. It reknit the complex ties of a society that had been interrupted for five years. It felt good to be back home, wrote Galomphete, "after such long time having been denied the opportunity by the government. We really enjoyed that moment."

I had been invited and would have very much liked to join them. Over seven years in southern Africa I had grown fond both of the Kalahari and of Botswana's predominantly decent people. To Botswana I had dragged my girlfriend, where we conceived a child, became engaged, and began to raise our daughter. When work ended and my family moved home to America, I left an off-road vehicle and camping equipment behind, with plans to visit again and again. And then, return became impossible.

After four decades as the exceptional shining example of African democracy, citizens became anxious their country was "heading for a dictatorship." Even the jovial bestselling novelist Alexander McCall Smith, creator of Precious Ramotswe, spoke up against Botswana's policies in the Kalahari. In response the government continued to label its critics as "enemies of the state." Church leaders grew "afraid of making public comments for fear of being accused of being members of opposition parties." President Mogae expelled a university lecturer as "a rogue" because he questioned the evictions of the Bushmen.[1] And on March 29, Mogae targeted seventeen critics—academics, human rights campaigners, and journalists from the United Kingdom, the United States, Australia, and Canada—and on that blacklist I found my own misspelled name.[2]

Naturally my initial reaction was pride: I was a freelance journalist thrust into the ranks of notable reporters and editors from the BBC, *Financial Times, Los Angeles Times,* and *Times* of London. I boasted to friends that I belonged to a club more exclusive than Nixon's Enemies' List. But my swagger softened to a larger sense of dismay. Short-term competition to commandeer and control precious water resources had driven an otherwise open and promising democracy down the slippery slope toward aggressive authoritarian rule. To be sure, on April 1, 2008, Botswana elected a new president, Ian Khama, whose government may yet mature. But reports of a "worrying trend" concerned how Botswana's "government responds to dissent" amid "perceptions it is becoming dictatorial and eroding its hard-fought record on democracy."[3] Indeed, Robert Mugabe of

Zimbabwe followed the example of Festus Mogae: destroying water supplies to punish dissidents, then expelling writers guilty of "committing journalism."[4] There has been no indication since that blacklist that Botswana will lift its visa restrictions against those of us it named.

Until that happens, I have these notes, photos, and memories. Shortly after the landmark ruling but before getting banned, I took one last journey into Botswana's dry heart with Qoroxloo's grandsons Galomphete and Smith, where we broke the news and lifted Metsiamenong into a festive mood. Qoroxloo's band recognized how, by never leaving the Kalahari as the siege constricted inward, they had ensured that Bushmen still had a home to which all might one day return. These happy few were proud; their core society had remained intact even as others melted away.

One evening at dusk, several of us walked a quarter mile to pay our respects before three unmarked piles of sand. The sky blazed orange and purple against thin distant clouds. Mongwegi pointed out where Qoroxloo lay buried among the honored elders. No animals had disturbed her grave over the last twelve months. The elements had smoothed the sand surface, but thorn branches still arced across it as a barrier to scavengers.

Jumanda Gakelebone shook his head. "In here are these three graves from five years," he said. "And during that time, how many dozens of young men and women have we buried outside the reserve? How many died out there and were unable to come home?"

Suddenly Mongwegi closed his eyes, threw back his head, and cried out "to the spirits of these ancestors" to "continue to guide the living. To give strength to us. To show us the way when we become confused and have doubts."

We turned in the silence and walked slowly toward the fires and the laughter.

On that same trip, I also camped in the government resettlement camp, Kaudwane, outside the reserve. Hundreds still lingered in purgatory, hating the place, but too fearful of government reprisals to leave. I arrived there with Roy Sesana and the FPK legal team, who reassured the Bushmen that the high court had ruled on five counts that they had been wrongfully, illegally, and unconstitutionally forced from the Kalahari and deprived of their livelihoods and homeland. They were free to return to the place they still called home. At this news the Bushmen had jumped up and danced and clapped excitedly, and their wide eyes shone in disbelief.

Later, as the initial joy subsided and reality set in, a few asked if the government would again provide them water inside.

The Bushmen's attorney, Gordon Bennett, slowly shook his head and looked down. On that count, he conceded, the high court had ruled termination and destruction of water was legal and constitutional. Botswana would not restore it. But he planned to negotiate a compromise with the government.

Upon translation, the Bushmen fell silent.

An hour after hearing the legal decision, two dozen Bushmen gathered beneath a shade tree, squatting on their haunches. As each spoke, the others listened. Some gestured, shook their heads, or drew in the sand with sticks.

I sat near a young Bushman, Kelejetseeing Moloreng. His family had raised him in Mothomelo, but he left the reserve as a teenager after officials destroyed the borehole. He now had a wife and child, and wanted to take them home in the Kalahari. I nodded at the discussion and made a questioning face.

"The men, they are talking," he said in the halting English he had picked up outside the Kalahari, "about water, how to get the water."

Two of his friends elaborated. Men were debating distances, the limits of mobility into the reserve, how much water they could carry, and what the government might allow. Women worried about the status and distribution of the wild food growing inside.

"Some we eat," explained Moloreng, "others we drink. They are divided. The food has water inside it."

I asked Moloreng and his friends if they planned to return home. All three vigorously nodded their heads, but their eyes left room for doubt. Some had previously been beaten for hunting. Most had been dependent on government services all their lives. "When it doesn't rain, it is a problem," said one.

"We grew up with it provided," added another, "and here was always a tap."

"But some of your people have never left the reserve," I observed.

"Yes," said Moloreng, looking away for a moment, and then back at me. "They are strong. We are young. We go in during the wet season, when it is green. But during the dry season . . ." His voice trailed off.

I scribbled his words down onto a yellow pad. He watched my

chicken-scratch, smiled, and asked what I was doing. For decades out in the Kalahari, Bushmen had grown accustomed to anthropologists and wildlife researchers working on dissertations, but in recent years the foreigners with cameras, recorders, and notepads had grown increasingly rare. Perhaps the romantic mystique and novelty were wearing off, and evicted Bushmen were becoming just like 40 million other deracinated people. I explained that I wrote about drought and the struggle over water, and how what unfolded here may foretell what occurs in nations beyond Botswana's borders.

The connection to people far away seemed to cheer him. "Yes," he said. "Put this into a book, so the world can know about us and what was done to us. And we can tell the story to our children."

"Where will you tell it to them? Here in Kaudwane?"

"No," he answered. "Inside. In the reserve. In our home."

"How will you live there?"

"The old," said Moloreng. "They know." He fell silent for so long I wasn't sure he would continue. Then he added in a quiet voice, "They know how to live without the water."

EPILOGUE

What Would Bushmen Do?

THE OLD MAY KNOW, BUT THEY ARE VANISHING faster than the vast wild places that forged their existence. And dying with them is their indigenous wisdom, their survival strategies about how humans can successfully adapt to our hotter, drier, and less forgiving world.

The rarity of these last free and autonomous Kalahari Bushmen can make them seem precious; it can lead us to romanticize them as Rousseau's "noble savage." Yet that, I think, would diminish Qoroxloo's life and death. She was neither savage nor superwoman but rather a pragmatist who knew exactly where she was and thus who she was. She knew how to live and when to die. She successfully resisted eviction armed with nothing more than the intimate knowledge of her community and her place. And her nobility emerged when the stress from nature's finite limits was compounded by official force, whereupon she did what any mother would do in such circumstances. Cut off from water, confined to a small perimeter, she denied herself week after week so that her family could endure. During her defiant existence Qoroxloo left a legacy as rich and potent as the rock art of her ancestors. Through perpetual drought and a protracted siege she revealed glimpses of a pragmatic Bushmen code of conduct.[1]

Every society lives by a social contract, whether the religious authoritarian "thou shalt not" laws of the Ten Commandments or the Koran or the secular democratic "government shalt not" protection of liberties under America's Bill of Rights. Perhaps because it remained unwritten, the Bushmen's code of conduct had the advantage of being simple, durable, flexible, and adaptable to aridity. Then again, it had to be. Without any hierarchical police enforcement, Kalahari lives were governed by voluntary and egalitarian interactions and by the authority vested in the desert itself.

Our challenge is to crack their code. This effort began more than a century ago in Cape Town, when the remarkable Wilhelm Bleek and Lucy Lloyd began to decipher the unusual glottal "click" language of Bushmen and discovered that beneath a seemingly simple and naive exterior ran complex and sophisticated systems of communication, orientation, dance, belief, habit, trade, mobility, ritual, relationships, and laws. From their early work it seemed the code—fusing together so many rich and layered stories—owed a great deal to the sheer necessity of surviving and enduring protracted drought.

In subsequent decades outside interest in the Bushmen waned, and they risked becoming no more than an obscure curiosity or historical footnote, until a scarred and troubled South African veteran of World War II named Laurens van der Post plunged into the central Kalahari on a quest to film and document "some pure remnant of the unique and almost vanished First People of my native land, the Bushmen."[2] He found them in the driest place and time of year, and returned from the desert a true believer. Transformed by humble people who seemed to have escaped carnage, war, despair, guilt, greed, pollution, racism, sexism, exploitation, social castes, and arbitrary injustice, van der Post devoted his life to proselytizing a new gospel of the near miraculous existence of those whom time forgot. Behold, he proclaimed to the industrialized world, Africa's hunter-gatherers were not to be pitied but glorified, for they offered us our last shot at salvation.

Van der Post opened a passage for thousands of apostles who followed. A desert pilgrimage became a kind of ritualized purification ceremony, as academics got themselves back to the garden.[3] Exposed to the raw elements and to a radically different existence, the industrial world's disillusioned victims found peace and redemption through the cleansing example of animal-skin-clad illiterates who ate locusts and wild honey in a forbidding land in which the transformative catalyst was the contact of water.

The nonliterate forager became something of a postmodern oracle. His or her code offered anthropological answers to various traumas and uncertainties. Bushmen seemed to help bring clarity to every issue that mattered, from gender roles and child care to the origins of science and the ecological limits of Mother Earth. Bushmen birth control methods informed global overpopulation debates, while their trance dances and herbal cures challenged Western medicine with new ways of healing. The primeval mind was discovered in Bushmen rock art, and their low-carb, gluten-free subsistence diet of meat, nuts, fruits, and roots inspired nutritionists to rethink how the affluent should eat.[4]

No religious cult took root among these argumentative artists and university academics. Yet when confronting today's political and moral issues, even atheists began to reformulate the Christian fundamentalist question to ask: What would Bushmen do?

In seeking out Bushmen to unlock a secure path through the age of permanent drought, I battled to get beyond lingering mythology. I am not immune to romantic sentiment. I want to believe, like Rousseau, that dissidents like Qoroxloo and indigenous people on every continent are innately one with nature, that they are more ethical than the rest of us, that they have a built-in tendency toward restraint which somehow makes Bushmen, as one colleague put it, "the original conservationists." By this reasoning the Kalahari Bushmen, if empowered, would not squander opportunity or trust. They would avoid the stupid, selfish, reckless, destructive, wasteful, divisive, and shortsighted mistakes that the rest make outside the reserve, especially regarding a resource as fragile, vital, and magical as water.

Over the years my inner skeptic persuaded me otherwise. I hope my lost naïveté reveals less a cynical pessimism than realistic hope based on universal egalitarianism. In any case, the unvarnished anthropological record of human nature shows that our species' behavior does not vary by race or ethnicity, only by externally imposed social, ethical, political, and physical constraints.[5] Without such limitations, each of us looks out for his or her personal interest—using others or exploiting the environment—until power corrupts us absolutely.

By now it should be all too clear that in a dry, hot world, water equals power.

So, given unlimited access to borehole technology and unlimited fuel to pump and deliver water, Bushmen would likely invite upon themselves unprecedented levels of trouble, depletion, and exploitation from which they would most likely never recover. Just like us. Elsewhere in the Kalahari, outside their reserve, I had seen such internally disruptive disputes already emerging. Some Bushmen bands blocked new wells (that might have threatened old rights) from being drilled; others pumped water to exhaustion in order to irrigate desert farm plots. Dozens of pumps have been abandoned after fights that arose from lack of clear ownership; other wells have been taken over by more powerful elites. Some were commandeered by a few families with habitat-destroying livestock, and, in another case, the traditional Bushmen ties were splintered by the granting of individualized rights over water points. Perhaps it is as dangerous to introduce abundant water where there was none as to remove it suddenly where there was little.

This is neither to blame Bushmen for behaving just like us, nor to advocate paternalistic care that we would ourselves reject. Bushmen are not naive; they can make their own decisions. If Qoroxloo's surviving band wants to pump and transport their own water on their own land to ensure their own autonomy in their own Kalahari reserve, a truly democratic government must surely let them. Scarcity will impose its own checks on them just as surely as it has with us.

The trouble with this small-government libertarian mind-set is that those checks can suddenly swing out of control and all too often become fatal. In increasingly hot, dry, and crowded landscapes, whether Botswana or the American West, the thirsty scramble for water can, and does, lead to chaos, conflict, environmental loss, and violence. One might argue that after centuries of being exploited and dispossessed, Bushmen deserve their own opportunity to exploit the earth and dispossess each other, but let's assume they seek the same security we all crave as we enter the age of permanent drought. To that end, Qoroxloo's legacy reveals a more positive and hopeful side to human nature in the face of scarcity.

If our competitive demand for scarce water drives us apart and escalates tensions, this same finite supply of freshwater is also itself what ultimately drags us back and binds us together. We may not like the rule of increasingly scarce water, but at the same time we cannot escape it. And Qoroxloo's band demonstrated how to embrace that reality. Her fundamental rule of adaptation was not to organize and mobilize physical resources to meet expanding human wants, but rather to organize human behavior and society around constraints imposed by diminishing physical resources.

To reiterate this book's thesis: We don't govern water; water governs us.

What does that principle mean in practice, for Qoroxloo's band and for us? Chapter by chapter, we have seen how the scarcity of water governed the outcomes of all vital decisions: who and why to trust; where and when to disperse; what to eat; how much to consume; which plants were burned for fuel, used for construction, or gathered to drink. A water-secure diet emphasized diversified, nutritious, drought-resistant, and moisture-rich permaculture over tastier, storable, transportable bulk food, and was harvested nearby at peak water-ripeness. Since tastier feedlot cattle could not survive droughts, the wise hunted desert-adapted game species whose juicy meat concentrated metabolic water. Health, sanitation, and medical decisions adroitly embraced aridity to convert waste into fertilizer, establish a buffer zone from disease vectors, and provide treatments from the concentrated oils of plants. Unrestricted liberty allowed dispersal to more abundant water resources, reducing ecological pressure and political stress; and rewarded each individual for drawing on his or her unique knowledge of water extraction from a diversified portfolio of strategies. Wildlife was not vertically ranked or segregated species by species but rather shared the increasingly arid landscape while horizontally competing for its water resources. Manufacture of luxurious vanity items encouraged competition and reserved water for more urgent needs. Trained from childhood to avoid evaporation and leaks, Qoroxloo's band developed their technology in the service of water, sealing it from the hungry sand and sheltering it from the thirsty sun.

Rivalry over scarce water resources has always existed, but against primal instincts toward zero-sum violence our interdependence encouraged voluntary exchanges among networks that efficiently spread out risks while rewarding conservation both within and between bands. Conservatives call these informal markets, while liberals see a reciprocal system of egalitarian barter. Regardless of ideology, such exchanges emerge only when a society collaboratively agrees to define and defend a water resource that could be divested. Rain belongs to everyone and everything, but Bushmen honored long-standing individual and group rights to water resources: a sip-well, a pan, a buried and labeled water canteen, a field of tsama melons, a grassy hunting territory favored by eland or gemsbok, a wild cluster of fruit or water-filled trees growing along a seep line.[6] Extending rights beyond kin to strangers not only reduced short-term hostility and resentment but also helped expand an informal safety net of grateful recipients—a reliable form of drought insurance.

We can adapt these underlying principles outside their reserve, to translate the Bushmen code of conduct into our industrial society. After all, whether it pulses between a competing heart and brain, sinks down in the shared aquifer beneath our fenced-off private property, or flows in the common currents that run along or across our walled-off borders, water has always remained, quite literally, the connective tissue that links and rules our fates. Only this magical glue makes us collaborate to endure drought conditions at every level. If we are to prevent dehydration, domestic strife, or degeneration into the ruthless Hobbesian/Darwinian scenario—recall those baboons around a water hole or those first colonists at Roanoke or Jamestown—and if we are to avoid testing the nightmare hypothesis of a transnational water war, then we need to derive a system like the one that for millennia has sustained people in the Kalahari.

Given the scale and complexity of our current political economy, what might this system look like? How do we obey water's rule? If Qoroxloo's band ran America's waterworks, what would Bushmen do?

Based on my reading of the evidence, they'd organize us around the measurable contours of the hydrological unit where we live: water known to exist within an aquifer or river basin. Then, within that unit, their code would secure the fundamental and minimal amount of freshwater required to keep each human healthy and alive. Some researchers peg this quantity at seven to thirteen potable gallons per day for drinking, cooking, and basic hygiene; others ratchet the amount up to one hundred gallons per person per day. Let's conservatively assume the upper limit, which still lies below America's comfortable average, and secure it as a fundamental human right, the kind that the Bushmen owned, recognized, and respected in others. The flip side of our individual endowment demands is that each person owns this water as an individual responsibility.

Human nature takes over from there. Confronted with finite limits imposed by drought and siege, the Bushmen code of conduct allows people to negotiate informally over the water resources they require, reaching out to partners with whom to exchange if and when they need more or less. People increased supply by efficiently reducing demands, and the benevolent result of their integrated informal right to water brought Bushmen into a relative state of social abundance.

This informal right may seem on the surface like what liberals vehemently demand from the United Nations, in which under a binding convention governments collectively hold federal water on behalf of the public, safe from the clutches of commerce.[7] If anything, Bushmen sought the opposite. It was not trade itself they feared, but the lack of secure access to the water resources they needed to trade in the first place. Government's primary role would then be to uphold their individual or band's right to access water—water that they already inherently owned and traded in reciprocal, lateral, and mutually beneficial exchanges. Defense of this kind of individually defined and divestible water right is a far cry from the enlightened paternalistic eco-socialism espoused by the so-called global water movement. It more accurately reinforces Justice Unity Dow's assertion that water does not belong to the government: It belongs to each of us.

Or it would if we had not already given it away. All of us growing up in cities and suburbs have surrendered both our right and our responsibility to water to state-run or -regulated institutions. Most of these command-and-control structures are now teetering on the brink of physical failure or institutional collapse. The left wants trillions invested to improve all creaky public waterworks. The right wants to privatize them. Ideology aside, it matters little whether our taps and pipes and sewers can be traced back to a government utility or a corporate venture if both operate as absolute top-down centralized monopolies that impose involuntary and uncompetitive rates and quality about which we cannot, by definition, negotiate. Public or private utilities are neither good nor evil; but right now they still remove all real incentives and accountability to conserve water efficiently, while making us dependent on aging infrastructure, political fecklessness, wasteful approaches, and unreliable supply in a radically changing climate.

In an era of permanent droughts, that is not a desirable place to be.

Like Qoroxloo's band, however, we can use our will and our cunning to reclaim what has always been rightfully ours. Government must ensure equitable delivery of water, but it need not be the institution that delivers it. In a free democratic society we can demand that water agencies restore and protect our inherent human right to water—say, the first one hundred gallons per day, owned by each of us—in return for our once again taking responsibility for using it wisely, free to truck, barter, and exchange any surplus water within that right that we manage each day to conserve.

In the spirit of Bushmen, we could demand water exchanges within aridity's authoritarian rule: in other words, unlimited markets within natural monopolies.[8]

Multiple benevolent side effects could emerge from this voluntary exchange of conserved water, within a confined and distributed network, among autonomous individual rights holders like you and me. First, we would recognize the true worth of our water in a radically new light, just as a new owner views the same house or car differently from when she rented it. Next, we would take steps to reduce demand. That freshwater we waste down the john? Never mind a dual flush; let's replace it with a dry sanitation urine diversion toilet. That bathwater? Keep the long, hot shower, but screw in a low-flow, high-pressure nozzle pulse spray and hook the drainage pipe to the backyard garden. Filter and reuse water in the dishwasher and kitchen sink. Through reduced demand for water, frugal utility customers—accustomed to frequent flier miles or cell phone units in those sectors—could earn and accumulate water credits they could give to friends and family members, or sell to profligate neighbors, squandering strangers, and wasteful businesses. Figuratively or literally, we'd be inclined to convert our umbrellas into basins, capturing rain instead of shedding it; homes might explore going "off the grid," to gather and harvest precipitation from tree canopies and rooftops, and seal it up in underground storage tanks. Much of this water technology already exists.[9] But strong incentives to reduce demand and boost sources of supply would spur the innovation of new appropriate water designs,[10] devolving authority to local and even familial levels.

The best part of this system is political. Right now splinter interests all pressure politicians to keep water rates low, build more dams, drain more wetlands, pump more deltas, expand storm drains and sewers, and plunder more aquifers. With individual water endowments, self-interest would incline us to lobby in the opposite direction. We would nudge governments to raise rates higher and across the board, to reward our efficiency, make the water we conserved worth more, drive us to more efficient exchanges, and restore substantially more leftover wild water back to all those endangered aquatic species.[11]

However small, local, and interpersonal in its origins, this translation of the Bushmen code of conduct could be replicated and scaled from the bottom up, from urban utilities to irrigation districts to international transboundary waters. By redefining water as an owned and tradable right that turns costly conflict into symbiotic cooperation, security analysts suggest that exchanges like those among Bushmen could alleviate national security tensions over border-crossing aquifers and streams from the Rio Grande and Colorado to the Great Lakes and Columbia, perhaps even in the Middle East.[12] In other words, landlocked Botswana could learn from the Bushmen living within its dry heart how to break the quiet siege imposed by rival African states surrounding it.

Seven years after my first hapless venture into the besieged Kalahari, as the globalized industrial world confronts the reality of its vanishing water, I keep asking myself: What would Bushmen do?

My interpretation may or may not accurately convey what the late Qoroxloo would have outlined, either for her resilient and humble band or for our far more rigid and profligate civilizations growing dry, hot, and thirsty outside the Kalahari. Then again, even while living, Qoroxloo never was one to lay down rules or dictate advice to friends and family, let alone foreign strangers like us. She didn't write a code of conduct. She lived it. As drought dragged on, she danced against the armed and unthinking forces closing in on her until finally, and on her own terms, she broke free.

When I think of the permanent drought we face in the years ahead, I like to picture Qoroxloo as last seen by her band of foragers: calm, defiant, and aware, striding purposefully across the hot dry Kalahari sands while quietly singing an ancient song to herself . . . and to anyone else who might care to listen.

Ke Ka Kaex X!o
(Thank You)

Notes

Any quotation or conclusion in the text that is not referenced in this note section comes from my primary research drawn from extensive travel to and living in Botswana from March 2002 through January 2007, after which I could not return to the country.

That same period covers the length of the high-court case, from which court transcripts and other documents provided thousands of pages of evidence, containing recorded testimony from dozens of people. The raw testimony had the advantage of being taken under oath and under penalty of perjury, yielding statements that sometimes ran counter to the original allegations and rationales put forth by interested parties campaigning for or against Bushmen relocations.

During those years I filed two dozen extensive reports for the Institute of Current World Affairs and other publications about the causes and consequences of water scarcity in the region and approaches to coping with drought. I also worked with foreign correspondents, environmental activists, journalists, tourism operators, wildlife researchers, anthropologists, and human rights campaigners within and outside of Botswana. I led investigative safaris into the central Kalahari; hired translators; and spoke with employees of key water and wildlife agencies and local government, the Southern African Development Community, and De Beers, recording more than 120 interviews.

I rely extensively on the published and unpublished work of respected scholars. That said, my most reliable and authoritative sources on arid land strategies and coping with water scarcity are individuals who often can neither read nor write.

Introduction: To the Heart of the Matter

1. Spencer Wells, *The Journey of Man*; Elizabeth Pennisi, "The First Language?" *Science* 303, no. 5662 (February 27, 2004): 1319–20; Nicholas Wade, "In Click Languages, an Echo of the Tongue of the Ancients," *New York Times*, March 18, 2003; George Silberbauer, *Hunter and Habitat*, pp. 39 and 289; James Schreeve, "The Greatest Journey," *National Geographic*, March 2006.

2. J. C. Prichard's nineteenth-century anthropological survey, titled *Researches into the Physical History of Mankind,* described Bushmen thus: "Human nature is nowhere seen in a more destitute and miserable condition," quoted in Peter Godwin, "Going Home Again," *National Geographic,* February 2001, p. 94.
3. Government of Botswana Web site, http://www.gov.bw/ (accessed March 2002).
4. P. J. Schoeman, *Hunters of the Desert Land,* p. 212.
5. George Silberbauer, *Report on the Bushmen Survey,* 1965; Sandy Gall, *Slaughter of the Innocent.*
6. His promise was by no means the last; according to *Affidavits filed in the High Court at Lobatse,* in May, June, and July 1996, government representatives assured the ambassadors of Sweden and the United States, the British high commissioner, the Norwegian chargé d'Affaires, and an official of the European delegation that "social services to people who wish to stay in the Reserve will not be discontinued."
7. R. K. Hitchcock, "We Are the First People: Land, Natural Resources and Identity in the Central Kalahari," *Botswana* 28, no. 4 (December 2002).
8. Robert Hitchcock, "Removals, Politics and Human Rights," *Cultural Survival Quarterly* 26, no. 1 (Spring 2002).
9. Suzanne Daley, "Botswana Is Pressing Bushmen to Leave Reserve," *New York Times,* July 14, 1996; the *Johannesburg Star,* June 19, 1997; James Suzman, "Kalahari Conundrums," *Before Farming.*
10. David Quammen, *Song of the Dodo,* 1997.
11. Alice Mogwe, "Who Was (T)Here First?" *Botswana Christian Council,* March 1992.
12. Departing U.S. ambassador John E. Lange at Botswana Press Club, August 28, 2002; UN Human Rights Commission, special rapporteur on indigenous peoples, Rodolfo Stavenhagen, called Bushmen victims of "discriminatory practices" being "dispossessed of their traditional lands" so that their "survival as a distinct people is endangered by official assimilationist policies," April 2, 2002.
13. Ralph Hazleton, "Diamonds: Forever or For Good? The Economic Impact of Diamonds in Southern Africa," *Occasional Paper* no. 3, the Diamonds and Human Security Project, March 2002.
14. "Culture Under Threat: Special Report on the San Bushmen." IRIN, March 5, 2004.
15. "Water: A Soluble Problem," *Economist,* July 19, 2008.
16. Charles J. Vörösmarty et al., "Global Water Resources: Vulnerability from Climate Change and Population Growth," *Science* 289, no. 5477 (July 14, 2000): 284–88.
17. National Assessment, *Climate Change Impacts on the United States: The potential consequences of climate variability and change,* US Global Change Research Program, Washington, D.C., 2000; Rowan Williams, "Gift of God," *Resurgence* no. 227, November/December 2004.
18. Andrea Thompson, "Man vs. Nature and the New Meaning of Drought," *LiveScience,* April 2007.

19. Richard Seager et al., "Model Projections of an Imminent Transition to a More Arid Climate in Southwestern North America," *Science* (April 5, 2007): "The research shows that there is a broad consensus amongst climate models that this region will dry in the 21st Century and that the transition to a more arid climate may already be underway." James Kanter and Andrew Revkin, "Scientists Detail Changes, Poles to Tropics," *New York Times*, April 7, 2007: "In its most detailed portrait of the effects of climate change driven by human activities, the United Nations International Panel on Climate Change predicted widening droughts in southern Europe and the Middle East, sub-Saharan Africa, the American Southwest and Mexico"; and *Climate Change 2007: The Physical Science Basis: Summary for Policymakers*, contribution of Working Group I to the Fourth Assessment Report of the Intergovernmental Panel on Climate Change, p. 3: "Projections from 19 climate groups around the world, using different models, show widespread agreement that southwestern North America, and the subtropics in general, are heading toward a climate even more arid than it is today."
20. This is the per capita daily average; arid-region inhabitants typically use more: each Colorado resident consumes 200 gallons and in Nevada, 340 gallons. By comparison, Europeans use 53 gallons; Africans consume on average 3 to 5 gallons.
21. UNDP, *Beyond Scarcity*, 2006.
22. Jerry Harkavy, Associated Press, November 12, 2005; for the complete story, please see Elizabeth Royte, *Bottlemania*, 2008.
23. "Mining Related Contaminants Persist in Some Appalachian Coal Regions," U.S. Geological Survey press release, December 28, 2006.
24. "The Great Plains: The Water May Run Out," *Economist*, February 9, 2006.
25. American Rivers, "Drought: A Vision of America's Future?" *Most Endangered Rivers Report*, 2003.
26. Traci Watson and Paul Overberg, "Rivers Down to Barest of Levels," *USA Today*, March 28, 2002.
27. Maude Barlow, *Blue Covenant*, 2007.
28. James Hansen et al., *Proceedings of the National Academy of Sciences* 103, no. 39 (July 31, 2006): 14288.
29. Jehangir Pocha, "China's Water Supply in Danger of Drying Up," *San Francisco Chronicle*, September 5, 2004.
30. Peter Brabeck-Letmathe, "Limits to Growth?" *World Economic Forum*, January 27 2007, summary notes; Peter Brabeck-Letmathe, "The Other Inconvenient Truth," *International Herald Tribune*, October 7, 2008; and "A Water Warning," *Economist*, The World in 2009 (special issue).
31. R. Maria Saleth and Ariel Dinar, *The Institutional Economics of Water: A Cross-Country Analysis of Institutions and Performance*, p. 57.
32. "Global Water Shortage Risk for Manufacturing," *Marsh Center for Global Risk*, September 6, 2007.

33. "Business and Water: Running Dry," *Economist*, August 21, 2008; Heather Cooley, Peter H. Gleick, and Gary Wolff, *Desalination, with a Grain of Salt*, Pacific Institute report, 2006.
34. Jon Gertner, "The Future Is Drying Up," *New York Times Magazine*, October 21, 2007.
35. William Finnegan, "Leasing the Rain," *New Yorker*, April 8, 2002; Meera Selva, "Death Toll Rises as Kenyan Tribes Battle over Water," *London Independent*, February 7, 2006; Christopher Reed, "Iraq: Paradise Lost?" *Harvard Magazine*, January 2005; "Marshes of Southern Iraq: One Third of Paradise," *Economist*, February 26, 2005; Edward Wong, "Marshes a Vengeful Hussein Drained Stir Again," *New York Times*, February 21, 2004; "Resurrecting Eden," *National Wildlife*, February 2004.
36. Jacques Leslie, "Running Dry: What Happens When the World Runs out of Fresh Water?" *Harper's*, August 2000.
37. UNDP, *Beyond Scarcity*, executive summary, 2006.
38. Peter Schwartz and Doug Randall, *An Abrupt Climate Change Scenario and Its Implications for US National Security*, briefing report commissioned by Andrew Marshall, Pentagon, October 2003.

Chapter 1: Kalahari Rivals

1. I chose this phonetic spelling over Xoroxo because Qoroxloo's grandchildren wrote that this was the way for her name to be pronounced.
2. The Western keyboard can express the epiglottal clicks of Bushmen in four ways:

   / Dental click in which the tip of the tongue is placed on the hard ridge behind the upper incisors. When the tongue is pulled away, a short, gentle sound is produced similar to the Englishman's *tsk tsk* when expressing disapproval.

   ! In this click, the tip of the tongue is pressed against the alveolar ridge (where the hard palate begins to curve up to the soft palate) to produce, when sharply released, a loud, popping sound.

   ≠ The front part of the tongue is pressed against the alveolar ridge. On release a sharp, flat snap is produced.

   // Lateral click, in which the front part of the tongue remains on the hard palate while air vibrates at the sides.

   Strictly speaking, the Bushmen would be known as the "G/wi" and the "G//ana." After consideration I have chosen to betray this more accurate grammar, not in disrespect for the language complexity of the people but to make their story more accessible and less distracting for the reader.

3. High-court ruling, December 13, 2006.
4. *Newsnight*, BBC Documentary, July 24, 2005.
5. Asit K. Biswas, *Management of International Water: Problems and Perspective* (Paris: UNESCO, 1993), p. 142.
6. www.ruaf.org/node/179.
7. "Botswana: Africa's Prize Democracy; Why Is an Arid, Landlocked Country in Southern Africa Such a Success?" *Economist*, November 6, 2004.
8. Peter J. Schraeder, "Continuity and Change in U.S. Foreign Policy toward Southern Africa: Assessing the Clinton Administration," *Nordic Journal of African Studies* 10, no. 2 (2001): 131–47.
9. "Botswana: The Southern Star," *Economist*, March 27, 2008.
10. John D. Holm, "Diamonds and Distorted Development in Botswana," *CSIS Africa Policy Forum*, January 8, 2007.
11. Thomas Tlou and Alec Campbell, *History of Botswana*, 1984.
12. "Diamonds for Development, President Festus G. Mogae," *Rush Transcript*, Council on Foreign Relations, moderated by Ambassador Richard C. Holbrooke, October 10, 2006.
13. Sidsel Saugestad, "Improving Their Lives: State Policies and San Resistance in Botswana," *Before Farming*, April 2005.
14. Louis Nchindo, quoted in Botswana's *Mmegi Monitor*, March 5–12, 2002.
15. Robert Hitchcock, *Kalahari Communities: Bushmen and the Politics of the Environment in Southern Africa*, Copenhagen, IWGIA document no. 79, 1996, p. 23.

Chapter 2: Crossing the Threshold

1. Joram /Useb, "'One Chief Is Enough!' Understanding San Traditional Authorities in the Namibian Context," *WIMSA*, May 25, 2000.
2. Roy Sesana, remarks in acceptance of Right Livelihood Award, on behalf of First People of the Kalahari.
3. Saugestad, "Improving Their Lives."
4. "Botswana Cuts Bushmen Services," BBC, January 23, 2002.
5. *Hydrosocial contract*, coined in 1999 by Tony Turton, is a notion that elegantly captures the Hobbesian/Lockean role of water control in the formation of modern nation-states. See chapter 3, "The Hydrosocial Contract and Its Manifestation in Society: A South African Case Study," in *Hydropolitics in the Developing World*, ed. Anthony Turton and Roland Henwood, 2002.
6. Peter Stark, *Last Breath*, 2001, p. 253.
7. W. J. McGee, "Desert Thirst as Disease," *Interstate Medical Journal* 13 (1906), as quoted in A. V. Wolfe, *Thirst: Physiology of the Urge to Drink and Problems of Water Lack* (Springfield, IL: Charles C. Thomas, 1958).

Chapter 3: Intransigent Eve

1. Silberbauer, *Hunter and Habitat*, p. 11; personal communication with the author elaborated on this anecdote; subsequent correspondence with other historians of Botswana, Cecil Rhodes, and the colonial era of southern Africa.
2. Sandy Gall, *Slaughter of the Innocent*, chapter 5.
3. Elizabeth Marshall Thomas, *The Old Way: A Story of the First People*, 2006, p. 203.
4. Silberbauer. *Hunter and Habitat*, p. 22.
5. John Marshall, "Death by Myth," part 5 in *A Kalahari Family*, series of documentary film productions, covering 1992–2000 and produced in 2003.
6. Mark Dowie. "Conservation Refugees," *Orion Magazine*, November/December 2005.
7. Robert J. Gordon, *The Bushmen Myth: The Making of a Namibian Underclass*, 1992.
8. Edwin Wilmsen, *Land Filled with Flies: A Political Economy of the Kalahari*, 1989.
9. Andrew Smith, "Ethnohistory and Archaeology of the Ju'/hoan Bushmen," *African Study Monographs*, supp. 22, March 15–25, 2001; Karim Sadr, "Hunter Gatherers and Herders of the Kalahari during the Late Holocene," chapter 11 in *Desert Peoples*, p. 209.
10. Michael Hennigan, "National Geographic and IBM's Genographic Project: Charting the Migratory History of the Human Species," *Finfacts*, February 18, 2007.
11. Gary Stix, "The Migration History of Humans: DNA Study Traces Human Origins Across the Continents," *Scientific American Magazine*, July 7, 2008.
12. Spencer Wells, *The Journey of Man*; James Shreeve, "The Greatest Journey," *National Geographic*, March 2006.
13. Christopher A. Scholz, Andrew S. Cohen, et al., "East African Megadroughts Between 135 and 75 Thousand Years Ago and Bearing on Early-Modern Human Origins," *Proceedings of the National Academy of Sciences* 104, no. 42 (October 16, 2007): 16416.
14. Nicholas Wade, "In Click Languages, an Echo of the Tongue of the Ancients," *New York Times*, March 18, 2003; Christine Fielder and Chris King, *Sexual Paradox*, 2006.
15. Michael Schirber, "Heat and Drought Cost the Most Billion Dollar Weather Disasters," *LiveScience*, January 30, 2005; on why drought is so deadly, see Brian Fagan, *The Long Summer*, 2004.
16. Elizabeth Marshall Thomas, *The Old Way*, p. 294.

Chapter 4: The Desiccation of Eden

1. John Hanks, "The Kalahari Desert," in *Wilderness: Earth's Last Wild Places*, ed. P. Gil, 2002, pp. 385–92.
2. "The Nature of Drylands: Diverse Ecosystems, Diverse Solutions," International Union for the Conservation of Nature, 2008.
3. Jeff Hecht, "Global Warming Stretches Subtropical Boundaries," *New Scientist*,

May 26, 2006; Thomas Reichler et al., "Enhanced Mid-Latitude Tropospheric Warming in Satellite Measurements," *Science* 312, no. 5777 (May 26, 2006): 1179.

4. Jacques Leslie, "The Last Empire: Can the World Survive China's Rush to Emulate the American Way of Life?" *Mother Jones*, January and February 2008.
5. J. M Prospero and P. J. Lamb, "African Droughts and Dust Transport to the Caribbean: Climate Change Implications," *Science* 302, no. 5647 (November 7, 2003): 1024–27.
6. Mark R. Jury, "Economic Impacts of Climate Variability in South Africa and Development of Resource Prediction Models," *Journal of Applied Meteorology* 41, no. 1 (January 2002): 46–55.
7. Aiguo Dai, Kevin E. Trenberth, and Taotao Qian, "A Global Dataset of Palmer Drought Severity Index for 1870–2002: Relationship with Soil Moisture and Effects of Surface Warming," *Journal of Hydrometeorology* 5 (December 2004): 1117–30; subsequently presented as "Drought's Growing Reach: NCAR Study Points to Global Warming as Key Factor," press release, National Center for Atmospheric Research, January 10, 2005.
8. Tim Flannery, *The Weather Makers*, p. 132.
9. Peter DeMenocal, "After Tomorrow: Climate Science and Political Reality," *Orion Magazine*, January–February 2005.
10. Brian Fagan, *The Long Summer: How Climate Changed Civilization*, 2004.
11. Edward R. Cook et al., "North American Drought: Reconstructions, Causes, and Consequences," *Earth-Science Reviews* 81 (2007): 93–134.
12. Flannery, *The Weather Makers*, p. 134.
13. Rodolfo Acuna-Soto, David W. Stahle, et al., "Megadrought and Megadeath in 16th Century Mexico Historical Review," *Emerging Infectious Diseases* 8, no. 4 (April 2002): 360.
14. Brian Fagan, *The Great Warming: Climate Change and the Rise and Fall of Civilizations*, 2008.
15. Shepard Krech, *The Ecological Indian: Myth and History*, 1999, p. 58.
16. David W. Stahle et al., "The Lost Colony and Jamestown Droughts," *Science* 280, no. 5363 (April 24, 1998): 564–67.
17. Ibid.; Jill Lepore, American Chronicles, "Our Town," *New Yorker*, April 2, 2007, p. 40.
18. Timothy Egan, "The First Domed City," *New York Times*, June 16, 2007.
19. Seth Borenstein, "Jet Stream Moving Slowly Northward," Associated Press, April 18, 2008.
20. Robert Kunzig, "Drying of the West," *National Geographic*, February 2008, p. 97.
21. James Lawrence Powell, *Dead Pool: Lake Powell, Global Warming, and the Future of Water in the West* (Berkeley: University of California Press, 2009).
22. Tom Englehardt, "America's Water War," *Salon*, November 19, 2007.
23. Porcha Johnson, "Experts to Offer Tips for Drought Relief," *Raleigh News and Observer*, January 23, 2008.
24. John Fuquay, "State Prepares for Prolonged Drought," *Raleigh News and Observer*, January 23, 2008.

25. "Drought Could Force Nuclear Shutdowns: Nuclear Plants Could Have to Shut Down or Scale Back Due to Drought," Associated Press, January 23, 2008.
26. Jenny Jarvie, "A Parched Southeast Urges Self-Restraint," *Los Angeles Times*, October 7, 2007.
27. Shaila Dewan and Brenda Goodman, "A Slow-Motion Response to Drought in US South: Last-Ditch Efforts Belie a History of Inaction," *International Herald Tribune*, October 24, 2007.
28. Michael Dudley, "Cities Abandoned? Mass Migrations? The Questions No One Is Asking about Drought," *World Environment*, PlanetCitizen.com, November 18, 2007.
29. Leonard Doyle, "The Big Thirst: The Great American Water Crisis; the US Drought Is Now So Acute that, in Some Southern Communities, the Water Supply Is Cut Off for 21 Hours a Day," *Independent* (UK), November 15, 2007.
30. Lynn Waddell and Arian Campo-Flores, "Dry—And Getting Drier: The Severe Drought Has Georgians Praying for Rain—and Battling with Their Neighbors," *Newsweek*, November 16, 2007.
31. Linda Greenhouse, "Justices Consider Dispute on Use of Potomac River," *New York Times*, October 8, 2003.
32. Jenny Jarvie, "Gov. to God: Send Rain!" *Los Angeles Times*, November 14, 2007.
33. Matthew Jones, "All 11 Hottest Years Were in Last 13: UK Met Office," Reuters, December 14, 2007.

Chapter 5: Besieged and Besieger

1. Zimbani Maundeni, Botswana's Democracy Research Project, quoted by the reporter Moabi Phia, "Democracy Under Fire in Africa's Model Pupil Botswana," Reuters, April 26, 2007.
2. Joseph Balise and Paul Richardson, "Botswana to Rule on Bushmen Removal from Gem Region," Bloomberg News, December 12, 2006.
3. Interview with Survival International; reaction to student protestors at London's School of Oriental and African Studies.
4. G. Simon Hark, quoting Michael Walzer in "The Siege of Iraq," *Austin American Statesman*, March 20, 2005.
5. Peter Apps, "Sri Lanka Rebels End Water Siege, but Fighting Over?" Reuters, August 8, 2006.
6. UN Report on Siege of Sarajevo; also Chuck Sudetic, "Small Miracle in a Siege: Safe Water for Sarajevo," *New York Times*, January 10, 1994.
7. In 1993 when David Koresh and the Branch Davidians holed up at the Ranch Apocalypse in Waco, Texas, federal agents eventually cut all water to the complex, forcing those inside to survive on rainwater and stockpiled U.S. Army "Meal, Ready-to-Eat" rations: Neil Rawles, *Inside Waco* (television documentary), HBO. In the Beslan hostage crisis of 2004, children and adults were denied water and grew so fatigued by thirst they could hardly flee the carnage.

8. Larry Collins and Dominique LaPierre, *O Jerusalem* (New York: Simon and Schuster, 1972).
9. "Gaza Siege Puts Public Health at Risk as Water and Sanitation Services Deteriorate," *Oxfam*, November 21, 2007.
10. Gerald Koeppel, *Water for Gotham*, p. 15.
11. U.S. State Department, *Annual Human Rights Report*, 2004, 2005, 2006.
12. More precisely, by June 1991 the country's water system would "suffer a slow decline," although "infectious diseases prevalence in major Iraqi urban areas targeted by coalition bombing (Baghdad, Basrah) undoubtedly has increased since the beginning of Desert Storm."
13. Thomas Nagy, "The Secret behind the Sanctions: How the U.S. Intentionally Destroyed Iraq's Water Supply," *Progressive*, September 2001.
14. The agency blamed civilian water-related deaths on Iraq's regime; Botswana's leaders blamed Bushmen thirst on their own intransigence: "The majority of those remaining were women held at ransom by their husbands," said one report.
15. "Military Panel: Climate Change Threatens U.S. National Security." Environment News Service, April 16, 2007.
16. Josephus, first-century Roman historian; Yigael Yadin, *Masada: Herod's Fortress and the Zealots' Last Stand*: Recent years have brought challenges to both accounts and interpretations of events, and Yadin's role as chief of staff to the Israel Defense Force may have swayed his findings. That said, extraordinary facts continue to be unearthed. A recent account shows that a date seed, dormant for two thousand years in Masada, has been brought back to life.

Chapter 6: The Rule of Water

1. Alfred Lord Tennyson, "In Memoriam A.H.H.," Canto 56.
2. Robert Wright, *Nonzero*.
3. R. F. Flint, *Glacial Quarternary Geology* (New York: John Wiley, 1971), p. 457, cited in Michael Main, *Kalahari: Life's Variety in Dune and Delta*, pp. 7–19.
4. Stephanie Hancock, "Climate Change, Human Pressure Shrink Lake Chad," Reuters, February 2, 2007; United States Geological Survey and NASA: Landsat Project, Lake Chad, Africa (accessed online February 22, 2008).
5. Main, *Kalahari*, p. 18.
6. Silberbauer, *Hunter and Habitat*, p. 96.
7. Gideon Louw and Mary Seely, *Ecology of Desert Organisms* (London: Longman, 1982).
8. Barry Lovegrove, *The Living Deserts of Southern Africa*, 1993.
9. Main, *Kalahari*, pp. 145–49.
10. Veronica Roodt, *Trees & Shrubs* and *Common Wildflowers*, both 1998.
11. Mary Seely, Tenebriod Beetles, in Skaife, *African Insect Life*.
12. Hanks, "The Kalahari Desert," in *Wilderness*, pp. 385–92.

13. Richard Despard Estes, *The Behavior Guide to African Mammals*, 1991; J. du P. Bothma, ed., *Game Ranch Management*, 1989.
14. Silberbauer, *Hunter and Habitat*, p. 261.
15. Ibid., p. 277; Polly Weissner, personal communication; Anthony Bannister, Peter Johnson, and Alf Wannenburgh, *The Bushmen*, 1979.
16. Ginger Thompson, "Bushmen Share Herbal Secret with Drugmaker for a Profit." *New York Times*, April 11, 2003.
17. Kenneth F. Kiple and Brian T. Higgins, "Mortality Caused by Dehydration during the Middle Passage," *Social Science History* 13, no. 4 (Winter 1989): 421–37
18. Harold E. Pashler, *Steven's Handbook of Experimental Psychology* (New York: John Wiley, 2002), p. 672.
19. E. F. Adolph, *Physiology of Man in the Desert* (New York: Hafner, 1947), pp. 226–40.

Chapter 7: Dispersal

1. Joseph Birdsell, "Some Predictions for the Pleistocene Based in Equilibrium Systems among Recent Hunter-Gatherers," in *Man the Hunter*, ed. Richard B. Lee and Irven DeVore, 1968, p. 11.
2. Hitchcock, *Kalahari Communities*, p. 28.
3. Arthur Albertson, personal communication; Draft Management Plan of Central Kalahari Game Reserve, 2002.
4. Robert K. Hitchcock, "Mobility Strategies Among Foragers and Part-Time Foragers in the Eastern and Northeastern Kalahari Desert, Botswana," University of Nebraska, unpublished.
5. Silberbauer, *Hunter and Habitat*, pp. 246–70.
6. Lorna Marshall, "Sharing, Talking, and Giving: Relief of Social Tensions among !Kung Bushmen," *Africa* 31, no. 3 (1961): 231–49.
7. Peter Veth, "Cycles of Aridity and Human Mobility Risk Minimization Among Late Pleistocene Foragers of the Western Desert, Australia," in Veth, ed., *Desert Peoples*.
8. Birdsell, "Predictions for the Pleistocene," 11.
9. Torsten Malmberg, "Water Rhythm and Territoriality," *Geografiska Annaler, Series B, Human Geography* 66, no. 2 (1984): 73–89.
10. Brian Wingefield, "America's Fastest Growing Counties," *Forbes*, March 22, 2007.
11. Asian Development Bank report on desertification, as quoted by Lu Tongjing in *Desert Witness: Images of Environmental Degradation in China* (Washington, D.C.: Heinrich Boll Foundation, 2003).
12. UN Report of the High Commission on Refugees; John Vidal, "Cost of Water Shortage: Civil Unrest, Mass Migration, and Economic Collapse: Analysts See Widespread Conflicts by 2015 but Pin Hopes on Technology and Better Management," *Guardian*, August 17, 2006.

13. David Bigman, quoting UN HABITAT, "Town & Country" letters, *Economist*. January 5, 2008.
14. Timothy Egan, "First Domed City," *New York Times*, June 16, 2007.
15. Xinyu Mei, "Why China Should Move the Capital," *Financial Times China*, December 27, 2007.

Chapter 8: Forage or Farm?

1. "The Nature of Drylands: Diverse Ecosystems, Diverse Solutions," International Union for the Conservation of Nature, 2008.
2. Ataman Aksoy and Francis Ng, "Who Are the Net Food Importing Countries?" World Bank Policy Research Working Paper no. 4457, January 1, 2008.
3. Donald G. McNeil Jr., "Malthus Redux: Is Doomsday Upon Us, Again?" *New York Times*, June 15, 2008.
4. Researchers dispute that gathering is a monotonous routine, countering that it involves the trained ability to distinguish among hundreds of edible, inedible, medicinal, and toxic species of plants at various stages of growth, demanding constant recalibration; this evolutionary trait is constantly resurfacing in modern studies of the Western brain.
5. Silberbauer, *Hunter and Habitat*, p. 49.
6. Ibid.
7. In this respect, she imitated a small, tunneling Kalahari rodent, the molerat. Because its lips are situated behind instead of in front of its teeth, the molerat is incapable of drinking. Like Bushmen, it eats corms and tubers as much for their water content as for nutritional value, and eats only what it needs, storing some but leaving bulk of the younger plants to recover and grow, so it can later return for another meal.
8. Bannister, Johnson, and Wannenburgh, *The Bushmen*.
9. Thomas, *The Old Way*, pp. 10–14, provides an eloquent meditation on the versatility and probable evolution of the digging stick.
10. "Anthropology: Marching Backwards; Not All Hunter-Gatherers Are Remnants of the Stone Age," *Economist*, April 16, 2005.
11. "What Is the Role of Climate Change? The Ten Most Frequently Asked Questions About the Rise in Food Prices," United Nations Food and Agriculture Organization, July 2008, www.fao.org/worldfoodsituation/wfs-faq.
12. Silberbauer, *Hunter and Habitat*, p. 274.
13. "Noble or Savage? The Era of the Hunter Gatherer was Not the Social and Environmental Eden that some suggest," *Economist*, December 22, 2007, p. 131.
14. UNDP, *Beyond Scarcity*; Gleick, *The World's Water*, 2002–2003; "Water: Doubling Financing to Double the Results," *Asian Development Bank Review*, December 2006–January 2007.

15. David Molden, Charlotte de Fraiture, Frank Rijsberman, "Water Scarcity: The Food Factor," *Issues in Science and Technology*, Summer 2007.
16. Tony Allan, "Water Resources in Semi-arid Regions: Real Deficits and Economically Invisible and Politically Silent Solutions," in *Hydropolitics in the Developing World*, ed. Turton and Henwood.
17. Fred Pearce, *When the Rivers Run Dry*, 2006, p. 4.
18. Ministry of Agriculture, "National Master Plan for the Arable Agriculture and Dairy Development," Government of Botswana, 2002; T. F. Stephens, *Botswana. National irrigation policy and strategy—Irrigation situation analysis*, FAO Report, November 2003; Anne Moorhead, "Country Profile: Botswana," *New Agriculturist*, November 1, 2005. For irrigation water use see http://www.fao.org/nr/water/aquastat/water_use/irrwatuse.htm; for irrigation cropping patterns see http://www.fao.org/nr/water/aquastat/water_use/croppat.htm.
19. David Molden, ed., *Water for Food; Water for Life: A Comprehensive Assessment of Water Management in Agriculture*, 2007.
20. Jeffrey D. Sachs, "Act Now, Eat Later," *Time*, May 5, 2008.
21. Molden et al., "Water Scarcity: "Biofuels: The Promise and the Risks," briefing paper for World Economic Forum, January 2008.
22. Peter Brabeck-Letmathe, "The Water Crisis: Another Inconvenient Truth," *International Herald Tribune*, October 6, 2008.
23. Ibid.
24. "Water for Farming: Running Dry; the World Has a Water Shortage, not a Food Shortage," *Economist*, September 18, 2008.
25. Molden et al., "Water Scarcity: The Food Factor," shows five options for getting more food; these are echoed and expanded in some detail in the recommendations of SIWI, IUCN, IWMI, and IFPRI. *Let It Reign: The New Water Paradigm for Global Food Security*: Final Report to the Commission on Sustainable Development, Stockholm International Water Institute, Stockholm, 2005; and in Molden, *Water for Food*.

Chapter 9: Quest for Meat

1. "Knee-Clicking Antelope Flexes Pecs." Zoological site http://www.labspaces.net/93070/ (accessed November 12, 2008).
2. Isak Dinesen, *Out of Africa*, 1937.
3. Louis Liebenberg, *The Art of Tracking*, 1990, pp. 4–8.
4. Bothma, *Game Ranch Management*.
5. Lorna Marshall, *The !Kung of Nyae Nyae*, 1976.; Lorna Marshall, "N!ow," *Africa: Journal of the International African Institute* 27, no. 3 (July 1956).
6. Thomas, *The Old Way*, pp. 264–66.
7. Megan Biesele, *Women Like Meat*, 1993, p. 41.

8. Ibid., p. 42.
9. Marshall, *The !Kung of Nyae Nyae*, p. 270, cited by Biesele.
10. *The Great Dance: A Hunter's Story*, documentary feature by Craig and Damon Foster, Aardvark/Earthrise/Liquid Pictures/Off the Fence Production in association with KirchMedia, e.tv, and Primedia Pictures, 2001.
11. BBC *Newsnight*, Documentary, July 24, 2005; *Voice*, July 1, 2005.
12. On July 22, 2005, Dr. Jeff Ramsay, President Mogae's press secretary, put out a lengthy memorandum explaining how the president's response to documented allegations of Bushmen abuse was to send out his own armed officials to round up suspected Bushmen poachers one by one and separately interview them about whether they might wish to reconsider their public statements to the press. To no one's surprise, after a few hours alone in the Kalahari, they did. The ad-hoc "investigation committee" got Qoroxloo's brother to deny "having ever been either handcuffed to the bull bar of a Toyota Land Cruiser or 'trussed like a chicken' or having been made to run for a long distance." Under similar circumstances two others cracked and recanted being "tortured" or "ever having been abused in his private parts by any wildlife officers." Since these interrogations were apparently not recorded, nor witnessed by neutral parties, there is no way of knowing what might have provoked such adamant denials of what they had previously offered to the public, freely and on the record.
13. Presentation, Survey of Community Conservancies, *Namibian Nature Foundation*, 2003.
14. "The Last Places on Earth," *New Scientist*, December 2007.
15. Jared Diamond, *Guns, Germs, and Steel*, 1999, p. 164.
16. Paramahansa Yogananda, *The Autobiography of a Yogi* (Los Angeles: Self-Realization Fellowship, 1946), p. 483.
17. John Reader, *Africa: Biography of a Continent*, p. 116.
18. Skaife, *African Insect Life*, pp. 159–60.
19. Fred Pearce, "Botswana: Enclosing for Beef," *Ecologist* 23, no. 1 (1993): 25–29.
20. Marty Matlock, "Water Profile of Botswana," United Nations Food and Agriculture Organization, 2007.
21. "USTDA Grant Supports Borehole Rehabilitation in Botswana," report, Ministry of Finance and Development, August 9, 2006.
22. Main, *Kalahari*, pp. 58–62.
23. Guy Oliver, "Botswana: Where Cattle Have More Value than People or Heritage: The Clearing of Kalahari Game Reserve for Ranching and the Erection of Fences Hundreds of Kilometers Long Are Endangering the Bushmen and a Unique African Ecosystem," *Sunday Independent* (London), August 3, 1997.
24. John Reader, *Africa: A Biography*, 1997, p. 117.
25. Main, *Kalahari*, p. 59.
26. California's Water Education Foundation report, p. 238.
27. Marty Matlock, "Water Profile of Botswana," Food and Agriculture Organization,

in *Encyclopedia of Earth*, ed. Cutler J. Cleveland (Washington, D.C.: Environmental Information Coalition, National Council for Science and the Environment, June 28, 2007), http://www.eoearth.org/article/Water_profile_of_Botswana.

28. California Water Education Foundation report, p. 238.
29. BEDIA, "Report on Cattle," Botswana Meat Commission, 2007.
30. "Livestock's Long Shadow—Environmental Issues and Options." United Nations Food and Agriculture Organization, November 2006.
31. Bryan Walsh, "Skip the Steak," *Time*, March 26, 2007.
32. Christopher G. Davis and Biing-Hwan Lin, "Factors Affecting U.S. Beef Consumption," Outlook Report no. LDPM13502, October 2005.
33. "The End of Cheap Food: Rising Prices Are a Threat to Many; They Also Present the World with an Enormous Opportunity," and "Briefing, Food Prices: Cheap No More," *Economist*, December 8, 2007.
34. *The North American Hunting Dilemma*, http://www.peachlandsportsmen.com.
35. Oren Dorell, "American Hunter Is a Vanishing Breed: States Worry Decline Could Cripple Funding of Conservancy Programs," *USA Today*, October 23, 2007.
36. U.S. Fish and Wildlife Service, National Survey of Fishing, Hunting and Wildlife Associated Recreation, 2006, p. 206.
37. This reversal has already begun in populated drylands: the Midwest is "re-wilding" the landscape with pronghorn and bison; Australian environmentalists advocate hunting kangaroo for food; and southern African ranchers are replacing their cattle with kudu; see also James Workman, "The Gamey Taste of Virtual Water: Recurring Drought Turns Cattle into Kudu," *Letters: JGW-20*, Institute of Current World Affairs, October 1, 2003.
38. Sue Lloyd Roberts, "Kalahari Bushmen's Battle to Return Home," *Newsnight*, BBC, July 25, 2005.
39. Thomas McGuane, "The Heart of the Game," in *The Best of Outside* (New York: Random House, 1997).

Chapter 10: Survival of the Driest

1. Botswana was taking measures to correct this shortcoming. My last visit into the CKGR I encountered campgrounds in the south with enclosed places to hang bucket showers and long-drop Porta-Johns.
2. Watt and Breyer-Brandwijk, 1962, cited in Roodt, *Trees & Shrubs*.
3. Marc Reisner, *Cadillac Desert: The American West and its Disappearing Water*, p. 458.
4. Roodt, *Trees & Shrubs*.
5. I. Hedberg and F. Staugard, "Traditional Medicinal Plants," *Traditional Medicine in Botswana* (Gaborone, Botswana: Ipelegeng, 1989).
6. Marjorie Shostak, *Nisa: The Life and Words of a !Kung Woman*, 1981.
7. Robert K. Hitchcock and Melissa Draper, "Health Issues Among the San of Western Botswana," unpublished paper.

8. Roodt, *Trees & Shrubs;* Western scientists have confirmed that its stems and leaves contained volatile and fixed oils, potent chemicals like 4 percent plubagin, a phenol that is a hydroxyl derivative of benzene.
9. Main, citing Lee and DeVore, *Kalahari Hunter Gatherers,* 1976, p. 170: Three decades earlier, another Bushman further west in the Kalahari accidentally cut his foot on a poisoned arrow. The small amount of poison festered and spread until the man amputated his own foot, recovered from self-surgery and limped about on a homemade crutch. Another survived an unarmed fight with a leopard, face paralyzed and arms and fingers permanently damaged.
10. People's Health Movement/Medact, *Global Health Watch, 2005–2006: An Alternative World Health Report* (London: Global Equity Gauge Alliance and Zed Books, 2005).
11. Meera Selva, "Botswana and Its Bushmen," *Independent* (UK), September 9, 2005.
12. As told to Stephen Corry, Survival International, recorded interview, 2004.
13. "Health of Indigenous People in Africa," *Lancet* 367 (June 10, 2006): 1943.
14. UNDP, *Beyond Scarcity;* "Water Hazard: Clean Up—or Risk Many Deaths," *Economist,* March 19, 1998.
15. UNDP, *Beyond Scarcity;* Peter H. Gleick, "Dirty Water: Estimated Deaths from Water-Related Disease, 2000–2020," Pacific Institute, August 15, 2002.
16. To halve, by 2015, the proportion of people who are unable to reach or to afford safe drinking water.
17. UNICEF and WHO figures, quoted in *New York Times,* August 27, 2004.
18. Gleick, "Dirty Water."
19. Katherine Ashenburg, *The Dirt on Clean: An Unsanitized History* (New York: North Point Press: 2007).
20. John Leland, "Yes, There's Such a Thing as Too Clean," *New York Times,* August 31, 2000.
21. Diamond, *Guns, Germs, and Steel,* p. 205.
22. William Jobin, *Dams and Disease: Ecological Design and Health Impact of Large Dams and Irrigation Systems,* 1999, p. 5.
23. Ibid.
24. World Commission on Dams, *Dams and Development,* 2000.
25. First People of the Kalahari, statement, March 17, 2005.
26. Nthomang, "Relentless Colonialism: The Case of the Remote Area Development Programme (RADP) and the Basarwa in Botswana," *Journal of Modern African Studies* 42, no. 3 (2004): 415–35.
27. Hitchcock and Draper, "Health Issues."
28. Botswana government, "FPK Allegations Baseless—Ministry," *Gaborone Daily News,* March 23, 2006.
29. Silberbauer, *Hunter and Habitat,* p. 58.
30. Tumelo Sebelegangwana, age twenty-one, had been driven from the Kalahari reserve as a teen; she had three children outside the reserve and took on two more when her sister died; all five were orphaned when AIDS killed Tumelo. I

am sick now, she told researchers. I am about to die. "We were the first people from Molapo to be evicted. Here in New Xade there are different kinds of diseases that we do not recognize."

Chapter 11: Water for Elephants Only

1. Apologists for Botswana with whom I spoke maintained that such an intense focus by outsiders on a few thousand Bushmen in the central Kalahari takes the spotlight off the plight of the vast majority outside of the reserve. Fair point. But their fates are closely intertwined; the failure or success of the Kalahari dissidents would respectively weaken or enhance the rights, esteem, and political clout of all Bushmen throughout Botswana.
2. Hitchcock, in "Mobility Strategies," writes: "There were situations where elephants had died or been killed. In these cases, large numbers of people moved to those places in order to take advantage of the windfall. People would camp some distance away from the elephant carcass and proceed to butcher it, cutting the meat into strips which they would hang in the trees or on specially designed racks to dry. The meat was then taken back to their residential locations on the river, often with the aid of donkeys. It is interesting to note that in these cases, territorial rules were often relaxed."
3. The fate of this male mirrored the Bushmen hunters. For millennia he lived on average sixty-five years and once roamed southern Africa, free and fearless of enemies. But in recent centuries farmers and ranchers fenced him in, shrinking access to foraging grounds and water holes, shooting him as a trespasser on his ancestral land. Doting females watched over his childhood. In adolescence he practiced skills that could win him a mating opportunity. But first the matriarchs drove him off to learn from bachelors and reduce odds of inbreeding. He rapidly grew a large, convoluted temporal brain lobe that rivaled primates for brain-to-body-mass ratio; in scent capacity and memory he surpassed all. His moods swung from boredom to play to rage, lust, and frustration. He could hurl stones at rivals. He showed compassion and sacrifice and cooperated on tasks. He could recognize his reflection and display vanity, or supposedly "uniquely human" self-consciousness. He communicated through unique, complex, and creative means of touch, smell, and vocalization. As an elder, he was toothless from a gritty diet; apprentices looked after him in exchange for survival skills: namely, where to find water. When death came, some might return and linger for hours, contemplating his bones like Hamlet holding Yorick's skull.
4. Reader, *Africa: Biography of a Continent*, pp. 258–62.
5. Richard Leakey, *Wildlife Wars: My Battle to Save Kenya's Elephants*, 2002.
6. "DNA Tracking: Conservation a la Carte," *Economist*, March 3, 2007.
7. I. S. C. Parker, and A. D. Graham, "Elephant Decline: Downward Trends in

African Elephant Distribution and Numbers," 2 pts., *International Journal of Environmental Studies,* 34 (1989): 287–304 and 35 (1989): 13–26.

8. Lyall Watson, *Elephantoms: Tracking the Elephant,* 2002: The African elephant detects subterranean water by smell, by a built-in divining rod in tusks, by a sonar system based on the echoes of soft and sensitive feet as it walks, literally, on tiptoes. Its skin has no sweat glands but, astoundingly, does not need to perspire; it air-cools surface blood vessels like natural radiators and contains precious moisture inside, including an emergency water reserve near the throat.
9. "Special Report: Elephants & Us. How & Where They Live; Culling; Birth Control; The Great Kruger Debate; Conflict; Money; The Numbers Game; World Opinon; New Visions." *Africa Geographic,* April 2006.
10. Mark Dowie, "Conservation Refugees," *Orion,* November/December 2005, p. 71.
11. Executive summary, *Millennium Ecosystem Assessment,* millenniumassessment.org.
12. Ryan Shen-Hoover, "Botswana's Tourist Trade Roars," *Investing in Africa,* October 4, 2007.
13. Tom Price, "Defying Ban, Bushmen Return to Kalahari Homeland," *National Geographic Adventure,* July 26, 2004.
14. Lex Hes advocated this compromise as early as 1998 in "Bushmen: The Gods Could Be Crazy," Travel Africa Edition Five, Autumn 1998. It was also, theoretically, on the table during low-level discussions with the negotiating team, although there never seemed to be any real desire by the government to affirm this in practice.
15. Maitseo Bolaane, "Community-Based Wildlife Enterprise Development Models in Botswana: A Comparative Study of Khwai and Sankuyo in Ngamiland," unpublished paper.
16. For a thorough discussion of how and, to some extent, why Botswana differs so profoundly from its southern African neighbors in this regard, see Elizabeth Rihoy, "Devolution of Authority for Natural Resource Management: Is It the Answer and Is It Achievable?" unpublished dissertation, 2007.
17. In a cull, state-hired marksmen brain-shot the matriarch from a helicopter. Once dead, her extended family stuck close by, pressed against her fallen body, and could be more easily killed, one after the other, while screaming and emptying their bowels. The government stored and later sold off the tusks, distributing skins and meat to loyal voters. A breeding herd ceased to exist, except in the memories of other elephants.
18. Dowie, "Conservation Refugees."
19. Ibid.

Chapter 12: The Paradox of Bling

1. Adam Smith, *Inquiry into the Nature and Causes of the Wealth of Nations,* 1776, book I, chapter 4, no. 13.
2. Ibid.

3. Smith's possible (wrong) answer set the cornerstone of Marxist economic thinking that helped fuel so many revolutions spreading across postcolonial Africa taking place while Mogae was in school.
4. New jargon about "diminishing marginal utility" explained how after consuming a first essential priceless bucket of water (to drink) the useful value of water decreased; a second bucket (bathe) added little additional satisfaction; the third (clean dishes) less; the fourth (wash the car) less still, and so on. The formula applied to diamonds, too, but even after two or three extra diamonds, the marginal utility remained high, due to scarcity.
5. The paradox remained a mainstay of economist jokes: "If you are lost somewhere in the desert the marginal utility of water goes to infinity (and the marginal utility of diamonds may even be negative since you have to carry them with you)." Imagine *that* scenario.
6. The full story, masterfully reported by Edward Jay Epstein in *The Diamond Invention,* concluded: "De Beers had proved to be the most successful cartel arrangement in the annals of modern commerce. For more than a half century, while other commodities, such as gold, silver, copper, rubber and grains, fluctuated wildly in response to economic conditions, diamonds continued to advance upward in price each year. Indeed, the mechanism of the diamond invention seemed so superbly in control of prices—and unassailable—that even speculators began buying diamonds as a guard against the vagaries of inflation and recession. Like the romantic subjects of the advertising campaigns, they also assumed diamonds would increase in value forever."
7. Outside of southern Africa, diamond extraction was a mostly wet, sloppy process. One million autonomous miners scoured muddy streambeds in vast humid watersheds. Getting jewels to market led to bribery, corruption, and war. Hence "conflict," or "blood diamonds." Muddy watersheds weren't a risk in the Kalahari, where capital, labor, and product could be neatly organized. The geologically stable concentration allowed De Beers and Botswana to isolate their Promethean diamond production behind ten-foot-high electrified and barbed-wire fences, which eliminated smuggling and allowed processing on site. The only real cost of this total control was aridity.
8. David James Duncan, "How Much Gold is a River Worth?" *New York Times,* April 12, 1997.
9. William J. Cosgrove and Frank R. Rijsberman for the World Water Council, *World Water Vision: Making Water Everybody's Business* (London: Earthscan, 2000).
10. Aneme Malan, Statistics South Africa, http://unstats.un.org/unsd/envaccounting/ceea/PImeetings/WaterAccounting-SA.ppt#276,17,Comparison between countries (2).
11. Phillip Ball, *Life's Matrix: A Biography of Water,* 1999.
12. The rise of flawless synthetic diamonds accentuated a crack in the cartel's façade. By the late 1990s, diamond merchants mounted their first challenge against De Beers' monopolistic rough diamond distribution system. Elsewhere

in Africa, diamond miners revolted. As competitors proliferated, De Beers began losing its global grip. That decade it fell from producing nearly half the world's rough diamonds and selling 80 percent of its supply through London, to producing less than two-fifths and selling 45 percent. Hairline fractures invited closer scrutiny. The European Union investigated De Beers' monopoly, blocked acquisitions, and broke up its control of Russia's supply. The U.S. Justice Department renewed investigations into price-fixing; by 2002 the district court in New York ruled against De Beers on the grounds of antitrust. The UN cited De Beers as being involved in "conflict" or "blood" diamonds in the Congo.

13. Ernest Oppenheimer neatly described the company's philosophy in 1910, cited in Epstein, *The Diamond Invention.*
14. Janine Roberts, "Masters of Illusion." *Ecologist Magazine,* January 9, 2003: In the absence of competition, De Beers earned hundreds of millions and paid mine workers thirty dollars per week; diamonds extracted from Botswana cost it twelve dollars per carat.
15. Janine Roberts, *Glitter & Greed,* 2003.
16. Robert Glennon, *Water Follies,* 2002, p. 181: In America's dry states, that already had happened. In 1979 Shoshone Indians were given a dollar per acre for ancestral land occupied by white settlers, who then sold off mining rights to industries extracting two things: gold and water. One mine drilled seventy wells two feet in diameter and two thousand feet deep to pump thirty thousand gallons of water per minute. Other mines in Nevada's Humbolt Sink pumped 4–5 million acre-feet of water, lowering the groundwater table thirteen hundred feet. Mines dried up Shoshone sacred springs, homes of Bah-o-hah or "water spirits," in an act of "cultural genocide."

Chapter 13: Oriented Against the Sun

1. Robert K. Hitchcock, "Sharing the Land: Kalahari San Property Rights and Resource Management," unpublished paper, 2008.
2. Bannister, Johnson, and Wannenburgh, *The Bushmen.*
3. World Commission on Dams, *Dams and Development,* p. 8.
4. Speech by Roosevelt at the dedication of Hoover Dam, September 30, 1935, in *Franklin D. Roosevelt and Conservation, 1911–1945,* ed. Edgar B. Nixon, 2 vols (Ayer: New York, 1972), pp. 438–41.
5. Alfred Merriweather, *Desert Doctor,* 1969, p. 92.
6. McCully, *Silenced Rivers,* 2001; Probe International, "Earthquakes Caused by Dams: Reservoir-Triggered/Induced Seismicity," 2008.
7. Malcolm W. Browne, "Dams for Water Supply Are Altering the Earth's Orbit, Experts Say," *New York Times,* March 3, 1996, quoting Dr. Chao from Geophysical Research Papers.

8. World Commission on Dams, *Dams and Development*, p. 3.
9. Maarten de Wit and Jacek Stankiewicz, "Changes in Surface Water Supply Across Africa with Predicted Climate Change," *Science* 311, no. 5769 (March 31, 2006): 1917–21.
10. Ian Sample, "Forecast Shows Africa to Face River Crisis," *Guardian* (UK), March 3, 2006.
11. Syukuro Manabe et al., "Century-Scale Change in Water Availability: CO_2-Quadrupling Experiment," *Climatic Change* 64, nos. 1–2 (2004): 59–76.
12. Barry Nelson et al., "In Hot Water: Water Management Strategies to Weather the Effects of Global Warming," Natural Resources Defense Council, July 2007. The Colorado River basin was projected to shrink by half from 13 trillion gallons in 1950–99 to 7 trillion gallons in 2010–39; since 1950, spring runoff occurred one to four weeks earlier throughout the American West.
13. "Estimating Evaporation from Lake Mead," U.S. Geological Survey, Scientific Investigations Report, 2006, p. 5252.
14. Jon Gertner, "The Future Is Drying Up," *New York Times Magazine*, October 21, 2007.
15. James Workman, "What Goes Up Must Come Down," *Institute of Current World Affairs: Letters*, December 2003; McCully, *Silenced Rivers*.
16. Engineers saw evaporation control as the Holy Grail of water management; many chased it, some imagined they caught a tantalizing glimpse, but none brought home the elusive prize; Ian Scarman, "Evaporative Control Systems: A Case Study in Innovation," University of South Australia, 2003, http://www.cric.com.au/seaanz/resources/9ScarmanEvaporativeControlsystems.pdf.
17. David Stephenson, "Maximizing Reservoir Yield in an Arid Region," unpublished paper, 2005.
18. "Botswana National Water Master Plan," Snowy Mountain Engineering Corporation, 2005.
19. "Five Months to Go Before Gaborone Dam Dries," *Daily News* (Gaborone), May 16, 2005.

Chapter 14: Cradling Every Drop

1. James Workman, "Kalahari Earthsuckers: San Sip-Wells Versus Roughneck Rigs: Seducing Water from the Ground," *Institute of Current World Affairs: Letters*, September 1, 2003.
2. Barry Lovegrove, *The Living Deserts of Southern Africa*.
3. Anthony Traill, *A !Xoo Dictionary* (Cologne: Rudiger Koppe Verlag, 1994), p. 279; Patrick Dickens, *English–Ju|'hoan Ju|'hoan–English Dictionary*, 1994, p. 177; both cited by Robert Hitchcock in "San Water Usage in the Kalahari," unpublished paper, 2008.
4. According to Schaan, at $2.50 per 1,000 liters, Botswana's water rates rivaled

Las Vegas's, which meant a school that leaked 23,000 liters per hour lost $400,000 a year.

5. Alister Doyle, "Fixing Leaks Can Avert World Water Woes: Expert," Reuters, August 21, 2006.
6. "Water for farming: Running dry; The world has a water shortage, not a food shortage," *Economist*, September 18, 2008.
7. Peter Rogers, "Facing the Freshwater Crisis," *Scientific American*, July 23, 2008.
8. "US Leaks 6 billion gallons per day," United Press International, October 2, 2007.

Chapter 15: The Reckoning

1. P. K. Kenabatho and B. P. Parida, "Forecasting Water Demand for Effective Water Management During Drought Periods in the Greater Gaborone Area, Botswana," ActaPress, 2006, actapress.com/PaperInfo.aspx?PaperID=28249&reason=500.
2. Wame Selepeng, "WUC to Lose Revenue Due to Water Restrictions," *Mmegi/The Reporter* (Gaborone), December 16, 2004.
3. Tshepho Bogosing, "No Improvement in Water Situation," *Mmegi News*, April 7, 2005.
4. Lynn Waddell and Arian Campo-Flores, "Dry—and Getting Drier," *Newsweek*, November 16, 2007.
5. "Five months to go," *Daily News* (Gaborone), May 16, 2005.
6. Bogosing, "No Improvement."
7. Egan, "The First Domed City."
8. International Conference on Water Resources of Arid and Semi-Arid Regions of Africa (WRASRA), in a statement released by Roger Stephenson, conference organizer, University of Botswana.
9. Emmanuale Koro, "Dams in Arid Africa Increase Drought Risk," *Islam Online*, August 2, 2004.
10. Bryan Davies and Jenny Day, *Vanishing Waters*, 1998.

Chapter 16: Haggling over the Source of All Life

1. Susan Berfield, "There Will Be Water: T. Boone Pickens Thinks Water Is the New Oil—and He's Betting $100 Million That He's Right," *Business Week* cover story, June 12, 2008.
2. Roger Bate, "Use the Free Market to Solve China's Water Shortage," *Wall Street Journal*, August 20, 2004; "China's Water Supply: A Modest Proposal," *Economist*, October 26, 2006.

3. John Tagliabue, "As Multinationals Run the Taps, Anger Rises over Water for Profit," *New York Times*, August 26, 2002; in the 2008 James Bond movie, *Quantum of Solace*, our intrepid hero must stop a megalomaniac from cornering South America's water market for profit.
4. The Maoist Internationalist Movement cites Bushmen as proto-Marxists; http://www.etext.org/Politics/MIM/faq/hnature.html.
5. C. M Hann, *Socialism: Ideals, Ideologies and Local Practice* (London: Routledge, 1993), pp. 35–38.
6. See Barlow, *Blue Covenant*.
7. Lorna Marshall, "Sharing, Talking, and Giving: Relief of Social Tensions among !Kung Bushmen," *Africa* 31, no. 3 (1961): 231–49.
8. Silberbauer, *Hunter and Habitat*, pp. 234; Polly Wiessner, "Hunting, Healing and *hxaro* Exchange: A long-term perspective on !Kung large game hunting," *Evolution of Human Behavior* 23 (2002): 407–36.
9. Robert K. Hitchcock and Megan Biesele. "Rights to Land, Language and Political Representation: The Ju/'hoansi San of Namibia and Botswana at the Millennium," in *Tracing the Rainbow: Art and Life in Southern Africa*, ed. Stevan Eisenhofer (Linz, Austria: Arnoldsche Publishers, 2001).
10. Silberbauer, *Hunter and Habitat*, pp. 232–42.
11. Polly Wiessner, "Risk, Reciprocity and Social Influences on Kung San Economics," in *Politics and History in Band Societies*," ed. R. B. Leacock and R. B. Lee, 1982, pp. 61–84; Polly Wiessner, "Measuring the Impact of Social Ties on Nutritional Status among the !Kung San," *Social Science Information* 20 (1981): 641–78.
12. Polly Wiessner, "Banking Time and Banking Relationships: Perspectives from anthropology, ethology, and neuroscience," presentation at Time Banking Congress, Kingbridge, August 2004.
13. Alec Campbell et al., *Tsodilo Hills: Copper Bracelet of the Kalahari* (Ann Arbor, MI: University of Michigan Press, 2009), p. 32.
14. See Robert K. Hitchcock, *Kalahari Communities*, 1996; Biesele and Hitchcock, "Rights to Land," 2000.
15. Robert K. Hitchcock, "Mobility Strategies Among Foragers and Part-Time Foragers in the Eastern and Northeastern Kalahari Desert, Botswana," unpublished paper, University of Nebraska.
16. Turton and Henwood, *Hydropolitics in the Developing World*.
17. Douglas Jehl, "As Cities Move to Privatize Water, Atlanta Steps Back," *New York Times*, February 10, 2003.
18. The consulting firm Booz Allen Hamilton has projected that Canada and the United States will need to spend $3.6 trillion combined on their water systems over the next twenty-five years.
19. Given a baseline of one thousand Bushmen, independent observers estimated the costs of education, clinic, rations, and water at thirteen dollars per person per month. A District Council official estimated ten dollars per person; Qoroxloo cost a whopping thirty-three cents per day.

20. Charles C. Mann, "The Rise of Big Water," *Vanity Fair*, May 2007.
21. Koeppel, *Water for Gotham*, pp. 70–100.
22. Julie Madsen, "Getting Soaked: Corporations Seek Control of the Water Supply—Here and Around the World," *Utne Reader*, November–December 2002.
23. Jon Luoma, "The Water Thieves," *Ecologist*, January 3, 2004; Bill Marsden, "Cholera and the Age of the Water Barons," a yearlong investigation by the International Consortium of Investigative Journalists (ICU), Center for Public Integrity, February 12, 2003.
24. "Promoting Privatization," International Consortium of Investigative Journalists, February 3, 2003; Charles C. Mann, "The Rise of Big Water"; Barlow, *Blue Gold*; Snitow, Kaufman, and Fox, *Thirst*.
25. Dan Dorfman, "Thirsting for Water Stocks," *New York Sun*, June 20, 2007.
26. Ibid.
27. Shawn Tully, "Water, Water Everywhere," *Fortune*, May 15, 2000.
28. "Water Privatization Becomes a Signature Issue in Atlanta," International Consortium of Investigative Journalists, February 12, 2003; Jehl, "As Cities Move to Privatize Water, Atlanta Steps Back."
29. Jon Luoma, "The Water Thieves," *Ecologist*, January 3, 2004.
30. Jehl, "As Cities Move to Privatize Water, Atlanta Steps Back."
31. "Bolivia; Water, Oil and the Mob: Another Defeat for Privatization as the Sick Man of South America Becomes Even Harder to Govern." *Economist*, January 22, 2005.
32. Gleick et al., *The New Economy of Water*, 2002.
33. Snitow, Kaufman, and Fox, *Thirst*.

Chapter 17: Human Rights, Water Wrongs

1. "Country Reports on Human Rights Practices—2005," Bureau of Democracy, Human Rights, and Labor, March 8, 2006.
2. Backed by bipartisan congressional support, U.S officials were at that moment employing what a former CIA director described as the "professional interrogation technique" known as "waterboarding." Officials immobilized the hooded witness horizontal or upside down and repeatedly poured water onto his face; because he was convinced he was drowning, a gag reflex kicked in; he choked, sputtered, and cracked in fourteen seconds. The psychological effects lasted much longer; years later some traumatized victims couldn't take showers or panicked when it rained.
3. "CIA Whitewashing Torture: Statements by Goss Contradict U.S. Law and Practice," Human Rights Watch, November 21, 2005.
4. Richard Esposito, "CIA's Harsh Interrogation Techniques Described," ABC News, May 19, 2006.
5. Kathleen Dean Moore, "Life, Liberty, and . . . Water?: In the Struggle over Water, Human Rights and Environmental Ethics Flow Together," *Orion*, Winter 2002.

6. Maude Barlow and Tony Clarke, *Blue Gold*; Jeffrey Rothfelder, *Every Drop for Sale*; Snitow, Kaufman, and Fox, *Thirst*.
7. Peter H. Gleick et al., *The New Economy of Water*, 2002.
8. "The Treaty Initiative to Share and Protect the Global Water Commons." http://www.blueplanetproject.net/cms_publications/12.pdf.
9. Snitow, Kaufman, and Fox, *Thirst*.
10. The Treaty Initiative was unanimously endorsed by the eight hundred delegates from thirty-five countries to accompany demands on behalf of all the world's citizens, including, presumably, you.
11. Barlow and Clarke, *Blue Gold*, p. 239.
12. "Stand Up for Your Rights: The old Stuffy Ones, That Is: Newer Ones Are Distractions," *Economist*, March 24, 2007; "Many Rights, Some Wrong: The World's Biggest Human-Rights Organization Stretches Its Brand," *Economist*, March 24, 2007.
13. Reinout Wibier, *Economist*, letter to the editor, April 7, 2007.
14. The World Bank's water guru, John Briscoe, dismissed the movement's underlying premise: "What does it mean to say that water is a human right?" he demanded. "Those who proclaim it so would say that it is the obligation of the government X to provide free water to everybody. Well, that's a fantasy."
15. Gleick, "The Human Right to Water."
16. *Ditshwanelo* is pronounced *di - tsua* [silent *h*] - *ne - lo*, with *di* being the plural prefix and the accent being on the *lo*.
17. Mogwe, *Who Was (T)here First?*
18. Letters of Correspondence between Department of Wildlife and National Parks (Jan F. Broekhuis) and Amogelang Segootsane c/o Ditshwanelo, July 5, 2002.
19. Ibid.
20. Constitutions of Ethiopia, Gambia, South Africa, Uganda, Zambia, and Burkina Faso; Stephen C. McCaffrey, "The Human Right to Water Revisited," in *Water and International Economic Law*, ed. Edith Brown Weiss, Laurence Boisson DeChazournes, and Nathalie Bernasconi-Osterwalder (Oxford: Oxford University Press, 2004).
21. Dandonoli, "The Human Right to Safe Drinking Water."
22. United Nations Economic and Social Council, General Comment No. 15: The Right to Water, Geneva, November 26, 2002.
23. That is: "adopt effective measures to realize, without discrimination, the right to water." Botswana would "have a constant and continuing duty . . . to move as expeditiously and effectively as possible toward the realization of the right to water." Specifically, Botswana was obliged to: respect the right by refraining from unfairly interfering with people's access to water, for example, "disconnecting their water supply"; it had to protect people from interference with their access to water by others, for example, price increases no one could afford; and it had to fulfill the right by taking all steps—legislation, implementation, monitoring—with available resources to realize the right to water.

24. Dandonoli, "The Human Right to Safe Drinking Water."
25. Letters of Correspondence between Department of Wildlife and National Parks (Jan F. Broekhuis) and Amogelang Segootsane c/o Ditshwanelo, September 13, 2005.
26. Sir William Blackstone, *The Commentaries on the Laws of England*, 4th ed. (London: J. Murray, 1876), 33–34, cited by Robert F. Kennedy Jr. in foreword to Ken Midkiff, *Not a Drop to Drink*, 2007. Specific text reads: "All forests that have been created in our reign shall at once be disafforested. River-banks that have been enclosed in our reign shall be treated similarly." "All evil customs relating to forests and warrens, foresters, warreners, sheriffs and their servants, or river-banks and their wardens, are at once to be investigated in every county by twelve sworn knights of the county, and within forty days of their enquiry the evil customs are to be abolished completely and irrevocably. But we, or our chief justice if we are not in England, are first to be informed."
27. Jefferson's inalienable rights might have been endowed by a Creator, but they were meaningless until enforced with bullets against redcoats, then soaked in Civil War blood, and subsequently earned through aggressive demands, burned churches, cracked skulls, and martyrdom of female suffragettes and black civil rights activists, by two world wars and countless guerrilla skirmishes.

Chapter 18: Primal Instincts and the Realpolitik of Water

1. Don Hinrichsen, "A Human Thirst," *World Watch Magazine*, January–February 2003.
2. Thomas, *The Old Way*.
3. David and Carol Hughes, *Last Feast of the Crocodiles: Death and Survival in a Shrinking Pool, National Geographic* documentary film, 1995.
4. "Take Them to the River: Water Rows in the South; Georgia Opens a Northern Front in Its Battle with Drought," *Economist*, March 13, 2008: In parched desperation, Georgia opened up an aggressive new battlefront. Its politicians decided that the border with Tennessee, demarcated by James Camak in 1818 with a faulty sextant, was mistakenly drawn a mile too far south. Now 190 years later, they wanted to correct that error and move the disputed boundary back up. Doing so would absorb half of Chattanooga; but territory wasn't the target. The adjustment would put Georgia in possession of a strategic slice of the Tennessee River.
5. Pearce, *When the Rivers Run Dry*, p. 186.
6. Douglas Jehl, "In Race to Tap the Euphrates, the Upper Hand Is Upstream." *New York Times*, August 25, 2002.
7. http://findarticles.com/p/articles/mi_m1310/is_2001_Oct/ai_79560849.
8. Patricia Kameri-Mbote, "Water, Conflict and Cooperation," *Navigating Peace*, issue 4, Wilson Center, January 2007.

9. Ibid.
10. T. F. Homer-Dixon, "Environmental Scarcity, Mass Violence, and the Limits to Ingenuity," in *Current History* 95 (1996): 359–65; Gleick, *The World's Water*, 2002; Robert Kaplan, "The Coming Anarchy: How Scarcity, Crime, Overpopulation, Tribalism, and Disease are Rapidly Destroying the Social Fabric of our Planet," *Atlantic Monthly*, February 1994 (Kaplan interviewed, based his argument on, and fully credited the work of Thomas Homer-Dixon, above).
11. Peter Annin, *The Great Lakes Water Wars*, 2006; Shiva, *Water Wars: Privatization, Pollution, and Profit*, 2002; Ward, *Water Wars: Drought, Floods, Folly and the Politics of Thirst.*
12. See the water conflict chronology in Gleick, *The World's Water: The Biennial Report on Fresh Water Resources, 2002–2003.*
13. 1938, Chiang Kai-shek covered his retreat from the Japanese army by dynamiting flood control dikes of the Huang He River, bogging down the enemy's army while killing an estimated one million of his own citizens.
14. Early in World War II, Britain blew up German dams to cripple industry while drowning downstream civilians. This event, with heavy casualties suffered by Britain as well, was portrayed in heroic terms in the film *The Dambusters.*
15. Under U.S. command, NATO troops targeted drinking water infrastructure in the Balkans, and in the 1991 and 2003 bombings of Iraq.
16. In order to depopulate Zaghawa in Darfur, Sudan's mounted Janjaweed destroyed village hand pumps and poisoned wells by stuffing them with corpses, just as U.S. troops reportedly had done decades earlier in Vietnam.
17. Under the UN definition of a "justified war," self-defense of national interests included rivers, so the need to protect a dry nation's lifeblood provided a strategic rationale to attack in a just war.
18. Aaron T. Wolf, "Conflict and Cooperation Along International Waterways," *Water Policy* 1 (1998), pp. 251–65; and Aaron T. Wolf and Joshua Hamner, "Trends in Transboundary Water Disputes and Dispute Resolution," in *Water for Peace in the Middle East and Southern Africa* (Geneva: Green Cross International, 2000).
19. Bjorn Lomborg, *The Skeptical Environmentalist*, 2001, pp. 150–58.
20. Tim Weiner, "Water Crisis Grows into a Test of US Mexcio Relations," *New York Times*, May 24, 2002.
21. Stephen C. McCaffrey, "Water, Politics and International Law," chapter 8 in *Water in Crisis: A Guide to the World's Fresh Water Resources*, ed. Gleiek, pp. 92–104.
22. L. Bloomfield and G. Fitzgerald, *Boundary Waters: Problems of Canada and the US, The International Joint Commission, 1912–1958*, Carswell, Toronto; cited in ibid.
23. Aaron T. Wolf et al., "International River Basins of the World," *International Journal of Water Resources Development* 15, no. 4 (1999): 387–427.
24. John Vidal, "Cost of Water Shortage: Civil Unrest, Mass Migration and Economic Collapse: Analysts See Widespread Conflicts by 2015 but Pin Hopes on Technology and Better Management," *Guardian* (UK), August 17, 2006.
25. "Kenya: Riven Over a River; Tempers Flare in the Dry Season." *Economist*, January 29, 2005.

26. S. Stayanath, E. Miguel, and Ernest Sergenti, "Economic Shocks and Civil Conflict: An Instrumental Variables Approach," *Journal of Political Economy* 112, no. 4 (2004): 727; Malcom Potts and Thomas Hayden, *Sex and War: How Biology Explains Warfare and Terrorism and Offers a Path to a Safer World*, 2008.
27. Martin Meredith, *The State of Africa: A History of Fifty Years of Independence* (New York: Free Press, 2006), introduction.
28. Turton and Henwood, *Hydropolitics*.
29. Maarten de Wit, "Changes in Surface Water Supply Across Africa," *Science*, March 3, 2006.
30. As long as its rivals were hamstrung by domestic strife and civil wars, Botswana had guzzled as much as it pleased. As Zimbabwe raged against white settlers, Botswana quietly tapped the Limpopo. As Namibia swarmed with guerrilla warriors, Botswana drilled and pumped aquifers for farms and cattle. As South African cities burned and rioted against apartheid, Botswana diverted water for industry, mines, and its capital. As Angola laced its upstream tributaries with land mines, Botswana developed downstream the continent's most lucrative tourism industry.
31. The regional superpower had a history of aggressively deploying water as a coercive tactic. The apartheid regime had cut off water to fifty thousand rebellious blacks protesting their condition. Whites had also backed a coup in tiny but mountainous Lesotho in part to divert west-flowing Orange River headwaters back east toward its own capital. Even Nelson Mandela approached water aggressively. On the Orange River headwaters Madiba sent helicopters to fatally attack perceived threats to its Katse Dam at its source in Lesotho; near its mouth he denied Namibia an equitable share of the precious current, refusing to renegotiate the colonial border demarcation.
32. Truce broke out in Angola with the assassination of Jonas Savimbi. As long as he had remained alive, no one went near the war-torn Okavango headwaters, an area littered with 10 million land mines and a cold-war history of tactical water deployment. In 1975 troops siezed the Ruacana hydropower complex to control the region's water resources. A decade later Cuban and Angolan forces launched a land and air attack on Clueque Dam, cutting power and destroying the water pipeline to Owamboland. Water was poisoned by one hundred bodies dumped into four drinking wells in the center of the country. Battles broke out on the Kunene River to control Gove Dam. Mercifully the civil war ended. But in the ensuing peace Angola dusted off Portuguese blueprints and announced plans to build a cascade of dams and diversions threatening the entire subcontinent with economic instability and political insecurity.
33. Yolandi Groenewald, "Namibia Targets Okavango for Power," *Mail & Guardian*, November 2002; Ian Michler, "Power or the Okavango," *Africa Geographic*, March 2003; Lars Ramberg, "Problems Facing the Management of the Okavango Delta," Harry Oppenheimer Okavango Research Center, November 1997.
34. Rothfelder, *Every Drop for Sale*, p. 159.

35. At issue was a 1.3-square-mile island in the middle of the river. Namibia called it "Kasikili." Botswana called it "Sedudu." The island lacked mineral wealth. No oil, diamonds, gas, or gold. It was often submerged during the rainy season. But as local Namibians periodically poled over in their *mekoro,* set up temporary reed huts, and fished or grazed whatever sustenance they could, Botswana watched and grew livid at the trespassing poachers. As tensions mounted, acrid squabbling began, "with intermittent threats of military action, including formal military occupation of the island by the Botswana Defense Force." The confusion stemmed from the 1890 Treaty of Berlin, which defined Namibia's eastern boundary vaguely as "the middle of the main channel," but for a century the river had been in constant flux. Botswana submitted its claims to the International Court of Justice in The Hague, which considered the depth, width, relative volume, navigability, and riverbed profile; after three years it ruled in favor of Botswana. But Namibia was allowed to navigate the waters unimpeded. That welcome precedent provided what might be a happy ending of a peaceful process, except new islands were forming. They rose up as the rivers dwindled. Six other islands in the stream remained disputed on the watery borders, and as currents sank and meandered they reversed the basis for peaceful judicial decisions.

Chapter 19: Intimations of Genocide

1. "Diamond Giant De Beers Opens New Store Amid Protests over Eviction of Bushmen in Botswana," Rush Transcript Radio Broadcast, *Democracy Now,* June 23, 2005.
2. Raphael Lemkin, "Chapter IX: Genocide," in *Axis Rule in Occupied Europe: Laws of Occupation—Analysis of Government—Proposals for Redress* (1944; repr., New Brunswick, NJ: Lawbook Exchange, 2005), pp. 79–95; for an excellent profile of Lemkin, his historical context, and the evolving definition of Genocide see Samantha Powers, *A Problem from Hell.*
3. Louis Anthing, Magistrate of Namaqualand, quoted in Gall, *Slaughter of the Innocent,* pp. 98–102.
4. Robert Gordon, *The Bushmen Myth: Making of an African Underclass.* Gordon draws parallels between colonial treatment of Bushmen in Africa and Nazi treatment of Jews in Europe that he finds "uncanny."
5. B. Voigt, quoted in Gall, *Slaughter of the Innocent,* p. 130.
6. Hannah Arendt, *The Origins of Totalitarianism,* cited in Gall, *Slaughter of the Innocent,* pp. 113–14.
7. There is, to put it mildly, robust debate over how these preconditions are defined or proactively curtailed. During the Clinton administration Gregory Stanton, president of Genocide Watch, offered to the U.S. State Department a paper called "The Eight Stages of Genocide." M. Hassan Kakar argued that conventional definitions are too restrictive and proposed his own. In December 2008 a

bipartisan Prevention of Genocide Task Force offered its own standards. The "precondition checklist" put forth in this paragraph is a synthesis of these three sources.

8. The World Bank's Africa demographer John May noted the 292 density of people per square kilometer; James Fairhead of the London School of Oriental and African Studies found agricultural land hit four thousand dollars per hectare when most lived on five hundred dollars a year. Jared Diamond, in *Collapse*, makes the same conclusion. As does James K. Gasana, "Natural Resource Scarcity and Violence in Rwanda," in *Conserving the Peace: Resources Livelihoods and Security*, ed. R. Matthew, M. Halle, and J. Switzer (Gland: IUCN, 2002).
9. "Ban Warns Business on Looming Water Crisis," *Financial Times*, January 25, 2008; UN secretary-general's address on water, World Economic Forum, January 25, 2008.
10. Arthur Chaptman and Hugo Maaren, Water Research Commission; John Carter of the Department of Water Affairs and Forestry; see Davies and Day, *Vanishing Waters*.
11. Mogae retaliated by putting the highest-ranking diplomatic, legal, and military experts in charge of an escalating campaign against the unarmed dissidents. General Merafhe, a decorated war veteran sworn to defend his nation against all enemies foreign and domestic, apparently saw in Qoroxloo's attempt to secure water a clear and present danger and national security threat.
12. Israeli prime minister David Ben-Gurion wrote in 1973, "It is necessary that the water sources, upon which the Land depends, should not be outside the borders of the future Jewish homeland," quoted in Janet Bush, "Profits Pour Forth—but the World's Poor Are Being Trampled in the 'Blue Gold' Rush," *Independent* (UK), August 6, 2006. Others assert that control of water has been the underlying source of conflict. See esp. Vandana Shiva, *Water Wars*, pp. 73–74; Fred Pearce, *When the Rivers Run Dry*, pp. 155–73.
13. Robert Olson, "An Israeli-Kurdish conflict?" *Kurdistan Report*, no. 24, December 1996; Nimrod Raphaeli and R. Barducci, "Potential Water Conflicts in the Middle East," *Middle East Media Research Institute*, no. 367, July 4, 2007; Pete Harrison, "Iraq Calls for a Water Treaty to Avert Crisis," Reuters, August 23, 2007; Vandana Shiva, *Water Wars*, pp. 71–72.
14. Keith Schneider and C. T. Pope, "China, Tibet, and the strategic power of water," *Circle of Blue*, May 8, 2008.
15. "Sudan: Watermelons, Conflict and climate change," *Integrated Regional Information Networks*, United Nations, May 13, 2008.
16. I borrow this phrase from Arundhati Roy's *The Greater Common Good*, an angry and eloquent essay protesting the government of India's justification for allocating waters to a few favored constituencies at the cost of the destruction and dispersal of millions of marginalized, indigenous "Scheduled Castes" who, in reaction, deploy Gandhi's passive resistance tactics against the democratic state.
17. Sandy Gall, "The Bushmen of the Kalahari," *Ecologist*, September 1, 2003.

Chapter 20: Escalation of Terrorist Activity

1. Spencer Mogapi, "The Rise of Sidney Pilane: Mogae loses his innocence," *Sunday Standard* (Gaborone), December 1, 2007.
2. Statement of Pilane in high court, September 5, 2005.
3. Court testimony, September 15, 2005.
4. The day's events have been reconstructed from interviews with more than a dozen primary participants and witnesses; parts of the incident were filmed by Jumanda Gakelebone; an extremely helpful resource was provided by the eyewitness Kuela Kiema, "The Trap to Catch Roy Sesana," working manuscript, Gaborone, 2006; see also "Botswana Country Reports on Human Rights Practices—2005," U.S. State Department, Bureau of Democracy, Human Rights, and Labor, March 8, 2006.
5. The debilitating effects of this tear gas were supposed to be temporary, but U.S. reports associated it with long-term damage to lungs, heart, and liver; other studies linked tear gas with miscarriages, given its effect of changing chromosomes in mammalian cells. Exposure usually required medical attention.
6. Rubber bullets also sounded harmless enough, almost playful, like snapped rubber bands. In fact they were typically metal bullets coated by or encased in rubber that could penetrate the eye; medical research concluded they killed and maimed too often to be considered safe for crowd control. Botswana officials recently had grown fond of them. In Mothomelo armed scouts shot an eleven-year-old son of a Bushman they were interrogating, hospitalizing him.
7. Section 74, *Botswana Penal Code*, Government of Republic of Botswana: "When three or more persons assemble with the intent to commit an offence, or, being assembled with intent to carry out some common purpose, conduct themselves in a manner likely to cause persons in neighbourhood reasonably to fear that the persons so assembled will commit a breach of the peace, will by such assembly needlessly and without any reasonable occasion provoke other persons to commit a breach of the peace, they are an unlawful assembly."
8. C. J. M. Drake, "The Role of Ideology in Terrorists' Target Selection," *Terrorism and Political Violence* 10 (1998): 53–85.
9. "FBI Says al-Qaida after Water Supply: Memo Says bin Laden Backers Scoured Web for Attack Ideas," Reuters, January 31, 2002.
10. American Water Works Association, "Protecting our Water: Drinking Water Security in America after 9/11" in H. Court Young, *Understanding Water and Terrorism* (Golden, CO: Transmountain, 2003), p. 43.
11. Nero laced his enemies' water with cyanide; Confederate soldiers left dead horses in water supply ponds to slow Sherman's march to the sea; Japan contaminated the water supplies of eleven Chinese cities; Germans released sewage into a Bohemian reservoir to bring a local population to its knees.
12. Young. *Understanding Water and Terrorism*, p. 43.
13. Marc Reisner, *Cadillac Desert*, 1986, p. 93.

14. Rocky Barker, *Saving All the Parts* (New York: Island Press, 1993), p. 85; L. J. Davis, "From Boise to the Back of Beyond," *New York Times*, September 26, 1999.
15. James Workman, "What Goes Up Must Come Down," *Institute of Current World Affairs: Letters*, December 2003.
16. Jonathan Mowazer, "Censorship and the Bushmen," *Index on Censorship*, http://www.ifex.org/en/content/view/full/51580/.
17. Of particular concern was the Bushmen's ability to sway or gain support from the United Kingdom and the United States; Botswana's government might raise $335 million from America, but it didn't want FPK dipping into foreign funds.
18. World Service Interview, BBC, February 20, 2003.
19. Like all advocacy NGOs whose meaning and existence depend on gaining exposure for its agenda, Survival could err on the side of hype and overreach and selective focus. Simplistic press releases outlawed shades of gray. Roy Sesana's flight to campaign was always a "dramatic journey" of Bushmen "fighting for their existence." Never mind 6 billion other humans, it warned how "climate change" or "biofuels" threatened a few dozen tribes. And its multiple campaigns blurred in curious ways: At one point the Bushmen appealed to the British "black population" for support; at other times the isolated Ayoreo tribe of Paraguay somehow mysteriously found itself "reaching out from the forests of South America to the Kalahari desert" to support Bushmen.
20. In a diary excerpt Corry told of frustrated efforts to meet with Botswana's government in Gaborone, and how despite the official perception of Botswana versus Survival, he found, "on the contrary, virtually everyone we talked to acknowledged that the Bushmen should be allowed to live on their ancestral land in the CKGR, and were angry at the way the matter had been mishandled by government. We found that not only the Bushmen themselves, but also many ordinary Batswana, had a far deeper understanding of the issues than the government thinks—as well as far more sympathy towards the Bushman victims than the government would like. The government, I thought, was seriously letting its people down, all its people, not just the Bushmen. Like Gandhi, I believe that a civilisation and a government will be judged not on how it treats most of its citizens—which is very easy to get right—but rather on how it treats its minorities. I left Botswana with renewed love and affection for its people and their unfailing good humour and politeness and, with heavy heart, asked my colleagues in London to plan the next three-year phase of the campaign."
21. Paul Kenyon, "Row over the Bushmen 'Genocide,'" *BBC: Crossing Continents*, November 6, 2005, broadcast November 10, 2005.

Chapter 21: An Open Heart

1. Jonathan Clayton, "Bushmen Beg to Keep their Freedom," *Times* (London), September 12, 2005.

2. Veronica Molosiwa, "Why Lift Arms Against Unarmed Citizens?" *Mmegi*, September 29, 2005.
3. "The Plight of the Gana and Gwi Tribes of the Central Kalahari Game Reserve." *RETENG*, October 18, 2005; "Government Accused of Arrogance," *Monitor* (Gaborone), December 4, 2006.
4. Meera Selva, "Botswana and Its Bushmen: Diamonds in the Desert and Despair in the Kalahari," *Independent* (UK), September 9, 2005.
5. Sello Motseta, "Government of Botswana Restricts Movement in Kalahari Game Reserve," Associated Press, September 2, 2005.
6. "Botswana: Government Denies Claims of Ethnic Cleansing." *Integrated Regional Information Networks*, October 10, 2005.
7. Survival International, "Submission to UN Human Rights Council on Indigenous Peoples and Access to Water," *ECOSOC*, April 11, 2007.
8. Barnaby Phillips, "Bitter Dispute over Bushmen Lands," BBC News, November 24, 2005.
9. "Interview with Rupert Isaacson," *Living on Earth*, radio program, aired October 14, 2005.
10. Richard Katz, *Boiling Energy*, 1982, p. 34.
11. Silberbauer, *Hunter and Habitat*, p. 175.
12. "Extreme Weather Events Might Increase," the World Meteorological Organization, July 2, 2007; Ewen Callaway, "Death Map USA: Natural Disaster Hotspots Revealed," *New Scientist*, December 17, 2008.
13. Thomas Crampton, "No End in Sight in Europe's Heat Wave," *New York Times*, July 25, 2006, noting the fifteen thousand who died three years earlier.
14. By Jennifer Johnston, "Climate Change Prediction: Scotland Faces Thousands of Deaths from Heatwaves," *Scotland Sunday Herald*, January 15, 2006; quoting Professor Paul Wilkinson of the London School of Hygiene and Tropical Medicine.
15. Stark, *Last Breath*, p. 141.
16. Facing three hundred thousand HIV-positive people, Botswana had donors offering hundreds of millions of dollars, and pharmaceutical companies donating antiretroviral drugs. But it lacked the human and physical infrastructure to absorb the load: no medical school; all of its doctors were immigrants or had been trained abroad; children born healthy in the hospital that day would on average die in thirty-four years in the lowest life expectancy on Earth.
17. Silberbauer, *Hunter and Habitat*, p. 68: The Bushmen of course conducted their own autopsies or postmortems on mammals, as often and as carefully as possible. Anything learned could serve them on future hunts. Interestingly, they did not dissect members of their own clans who had died either slowly or suddenly. Their knowledge is imperfect, but given their illiteracy and crude tools, it displays far greater understanding of the human body than medieval surgeons of Europe.
18. Thomas, *The Old Way*.
19. Bill Hayes, *The Anatomist*, 2007.

20. This scenario was dramatically played out in Laurens van der Post's story of finally encountering Bushmen in *Lost World of the Kalahari*; Bannister, Johnson, and Wannenburgh agree, in *The Bushmen*; this may have only been in cases of extreme stress, for in *The Old Way*, Thomas asserts that the old were never left alone in their dying moments.

Chapter 22: Release

1. Silberbauer, *Hunter and Habitat*, 112.
2. Biesele in communication to David Lewis-Williams.
3. In 1873 a Bushman named Qing took J. M. Orpen to a southern African painting of what appeared to be bloody-nosed men manhandling a crude four-legged creature. But it was far more. Qing explained that the "rain's men of power were capturing rain's animal, which was believed to live in a waterhole. They led it to a parched place to kill it so its milk could fall like rain."
4. David Lewis-Williams and Thomas Dowson, *Images of Power* (Halfway House, South Africa: Southern Book Publishers, 1989), p. 54.
5. David Lewis-Williams and Thomas Dowson "Through the Veil," *South African Archaeological Bulletin*, 1990.
6. Silberbauer, *Hunter and Habitat*, p. 112.
7. Rachel Holmes, *The Hottentot Venus: The Life and Times of Saartjie Baartman (Born 1789–Buried 2002)* (London: Bloomsbury, 2007).
8. Darren Schuettler, "African Reburied at Home, 170 Years On," Reuters, October 5, 1998.
9. "Burial Delay Worries Government," *Daily News* (Gaborone), December 8, 2005; "Government Accuses Ex-CKGR Residents of Delaying the Burial of Duxee," GABS-FM, December 8, 2005.
10. Rebaone Odirile, "Stand-Off Over Duxee's Corpse," *Botswana Guardian*, December 9, 2005.
11. Spencer Mogapi, "De Beers Urges Botswana to Halt Bushmen Evictions," Reuters, December 5, 2005.
12. Mpho Sekute, "Let Bo-rra Duxee Burry [*sic*], Mourn Their Mother," *Botswana Guardian*, December 14, 2005.
13. "Duxee Is Buried," *Daily News* (Gaborone), BOPA, December 30, 2005.

Chapter 23: The Verdict

1. "Gabs Dam Goes Up to 27%," *Botswana Mirror*, February 14, 2006.
2. Dikarabo Ramadub, "The Rise and Rise of GC Dam," *Botswana Guardian*, March 3, 2006.
3. "Water Restrictions Here to Stay," *Botswana Gazette*, February 15, 2006.

4. "Waste Water Sales Fall," *Botswana Echo*, March 9, 2006; "Water Restrictions Lifted," *Botswana Mirror*, July 26, 2006.
5. Gordon Bennett, "Summary of Issues in the Bushmen's Court Case," September 2005.
6. Dow enumerated her relevant factual evidence, which included: Roy Sesana's residences shared by three wives and six children; the exact numbers of adults and children relocated from each settlement; which settlements had arisen and which had vanished; how many Bushmen remained; the origins of the reserve; its founders, rainfall, rules, fences, and human and animal residents.
7. Dow noted that "the presence of the police in an operation of this nature and size would not, in itself, be curious; what is curious though was the persistent denial by government that there was any police presence at all." Dow also wondered why the government described those who stayed to look after an ailing relative as using a ploy: "The question becomes why someone who is not under pressure to relocate would need a ploy to remain in the Reserve."
8. "Toora! How Basarwa Found Their Song," *Mmegi*, December 14, 2006.
9. "Basarwa Going Back to CKGR," *Mmegi*, December 14, 2006.
10. "Government Will Not Appeal CKGR Case," *Mmegi*, December 19, 2006.
11. One of Europe's most evocative stories, and later film, tells of Jean de Florette, an exuberant hunchback tax collector who inherited a dryland farm and discovered how to thrive within the harsh rhythms of the arid landscape. But a neighbor, seeking control of natural resources, quietly cemented up the source of a natural spring, dried up the farm, and drove its rightful heirs to ruin, eviction, and death. The entire village knew what occurred, and one even witnessed the sinister deed. But for years all obeyed a rural *omerta* against ugly strangers. Only when the village was also crippled by drought did they speak the truth. A similar dynamic emerged in Botswana.
12. "Bushman Land Comes Home: But Botswana Government Will Provide No Services," *Mail & Guardian*, December 15–20, 2006.

Chapter 24: The End of the Beginning

1. Dikarabo Ramadubu, "Professor Good Gone for Good," *Botswana Guardian*, July 29, 2005.
2. Keto Sewai, "Government Slaps Visa Restrictions on Critics," *Mmegi*, March 29, 2007: "The seventeen affected individuals are: Steven Corry, Mirriam Ross, Fiona Watson, Jonathan Mazower, Janie Workman, Jonathan Reed, David White, John Walsh, Oliver Duff, Karin Goodwin, Carol Midgley, and Jonathan Simpson—all from the UK. The listed Americans are: Rupert Isaacson, Eric Grossberg and Tom Price; while Ian Taylor is an Australian and Daniella Stor is Canadian. At press time, *Mmegi* had been able to establish that four of the

Britons—Corry, Ross, Watson and Mazower are all from Survival International (SI), government's well-known nemesis in the CKGR saga. Seven of the people in the list are journalists. They include Simpson, the respected world affairs editor with the British Broadcasting Corporation (BBC); *Financial Times* of London's South African correspondent, Reed and its African editor, White; Price, a highly respected American freelance journalist who often contributes to major publications such as the *Los Angeles Times*. Other journalists are: Duff (*Independent*—UK), Goodwin (*Sunday Times*—Scotland), and Midgley (The *Times*—UK). Taylor is an Australian academic who previously worked as a lecturer on African Affairs at the University of Botswana. He co-authored a critical paper entitled "Presidential Succession in Botswana: No model for Africa" with Professor Kenneth Good. Speculation is rife that that paper, which Good was unable to present, led to his unceremonious deportation from Botswana. American Isaacson is known to be with the Indigenous Land Rights Fund, while Grossberg is suspected to be associated with an organisation that deals with conflict-free diamond issues. At the time of going to press, *Mmegi* had not yet established what Stor, Workman and Walsh do" (all oddities of spelling, punctuation, and fact: *sic.*)

3. Siphamandla Zondi of South Africa's Institute for Global Dialogue; the UN's special rapporteur on indigenous peoples, Mexican Rodolfo Stavenhagen, was subsequently kept from the country.
4. Botswana instructed all state media—print, television, and radio—to use positive government statements when reporting on the Bushmen; told to be "patriots first and foremost," October 25, 2006.

Epilogue: What Would Bushmen Do?

1. In a seminal chapter of *A Sand County Almanac* called "The Land Ethic," Aldo Leopold eloquently argues how the "extension of ethics is actually a process in ecological evolution." The evolving process or "ethical sequence" rippled outward, more inclusively with time, from: (1) personal conduct codes like the Ten Commandments, which guide relationships between individuals; to (2) social conduct codes like the Golden Rule or U.S. Constitution which guide and govern the relationships between people and society; to (3) natural conduct codes, still emerging and undefined, which integrate humans with our complex life support system. Leopold focuses his analysis on this third category:

   > An ethic, ecologically, is a limitation on freedom of action in the struggle for existence. An ethic, philosophically, is a differentiation of social from antisocial conduct. These are two definitions of one thing. The thing has its origin in the tendency of interdependent individuals or groups to evolve modes of cooperation. The ecologist calls these symbioses. Politics and economics are advanced symbioses in which the original free-for-all

competition has been replaced, in part, by cooperative mechanisms with an ethical content.

2. Van der Post, *Lost World of the Kalahari*: The author caught glimpses of Bushmen at the edge of a swamp, saw signs of their existence near a spring, and marveled at their mysterious artwork on smooth mountains rising out of the sand. Finally, he discovered and interacted with a band at ancient sip-wells.
3. Modern interactions with Bushmen echoed ancient narratives between Gilgamesh and Enkidu or even Jesus and John the Baptist: At the desert edge the doubting man encountered the near equal he was once akin to and who showed what he might yet be. He recognized himself in what seemed a more wholesome incarnation. Death would come to both with certainty, but the spirit would transcend it.
4. Bushmen appear even to have converted Madison Avenue. One recent ad marvels at how, "with a series of simple clicking sounds," a Bushman communicates complex messages that the rest of us can replicate only through the purchase of Ricoh's high-end digital imaging technology.
5. Matt Ridley, "Ecology as Religion," in *The Origins of Virtue*, 1996; Robert Wright draws equally convincing conclusions at length in *The Moral Animal: Evolutionary Psychology and Everyday Life*, 1994.
6. Sometimes ownership rights bore marks as obvious as a notched arrow or a labeled water canteen; more often than not, the ownership information was conveyed orally among those who asked permission.
7. See Maude Barlow, *Blue Covenant: The Global Water Crisis and the Coming Battle for the Right to Water*, 2007. As a tireless activist leader and social advocate, Barlow deserves credit for putting the right to water on the global radar screen, but her antiprivatization ideology blinds her to the practical fallacy of what she seeks. An inalienable right to water, held as a public trust by the government, which no one can trade, is the equivalent of a right to vote, held in trust by the state, which no one can cast. It restricts individual liberty and diminishes social opportunity. Thus Barlow seeks to disenfranchise in the name of empowerment.
8. As James Saltzman observed in *Thirst: A Short History of Drinking Water*, there is ample precedent for this combination: A rights-based water management regime is clearly not a new idea. The right to thirst in Jewish and Islamic law, sharing norms in Africa and India, and the "always ask" custom among aborigines all depend on a universal norm of access to drinking water by right in times of need. The ancient Roman *Aqua Nomine Caesar* practice of free water was rights-based; it was a right of provision guaranteed by the emperor. Treating drinking water supply as a priced resource is by no means a new idea, either. The *vectigal*, a tax on the private consumption of water, funded operation of the Roman water system for centuries. Private water vendors underpinned much of New York and London's water supply through the nineteenth cen-

tury, and now supply London once more. Nor, finally, are these two identities mutually exclusive.

9. Heather Kinkade-Levario, *Design for Water: Rainwater Harvesting, Stormwater Catchment, and Alternate Water Reuse* (Vancouver, Canada: New Society, 2007); see also Utility Consumers' Action Network, "San Diego's Challenge of the Century" report, August 5, 2008.
10. While most can-do, high-tech Americans hear "water scarcity" and react almost instinctively with a call for massive large-scale desalination, that kind of high-tech default option is geographically limited to affluent coastal cities like Tampa or Santa Barbara. Even then it is ridiculously expensive, costing one hundred to one thousand times what conservation or treatment would cost. The industrial process itself is energy-intensive, burning carbon in ways that worsen climate change, or splitting atoms that can't be disposed of. And to produce freshwater the plants must pollute the sea, dumping a residual saline mess back into the sea. Worse, as Bushmen have shown, technology by itself does nothing to counter our inherent incentives toward waste, or our dependence, and can leave us even more vulnerable should it fail or be shut down. If necessity is the mother of invention, Bushmen have shown us that the real invention lies in social innovation.
11. The Bushmen's survival strategies have also shown why we might be suspicious of the current top-down environmental flow regimes, requirements, and regulations. Experts have their place, and I by no means consider myself anti-intellectual or anti-elitist. But anyone who has tried to set aside a certain amount of water "for nature" faces the same lack of political credibility as another who tries to set aside water "for extractive industries" or "for agriculture." Each indirectly represents a vaguely defined constituency; my particular special interests may or may not diverge from your own, but in any case we each seek bigger slices from what we assume to be an expanding pie. We want it all.
12. Franklin Fisher and Annette Huber-Lee, *Liquid Assets: An Economic Approach for Water Management and Conflict Resolution in the Middle East and Beyond* (Washington, D.C.: RFF Press, 2005).

Bibliography

Abbey, Edward. *Desert Solitaire: A Season in the Wilderness.* New York: Ballantine Books, 1968.

Acuna-Soto, Rodolfo, et al. "Megadrought and Megadeath in 16th Century Mexico." *Emerging Infectious Diseases* 8, no. 4 (April 2002).

African Commission on Human and Peoples' Rights. "Report of the Working Group on Indigenous Populations/Communities in Africa: Mission to the Republic of Botswana," June 15–23, 2005.

Albertson, Arthur. "The CKGR Negotiating Team Visit to CKGR, Sep 26–October 2." Unpublished report, 2002.

Anderson, Terry L., and Pamela Snyder. *Water Markets: Priming the Invisible Pump.* Washington, DC: Cato Institute, 1997.

Annin, Peter. *The Great Lakes Water Wars.* Washington, DC: Island Press, 2006.

Arike, Ando. "Owning the Weather: The Ugly Politics of the Pathetic Fallacy." *Harper's,* January 2006.

Asmal, Kader. "Water as a metaphor for governance: Issues in the management of water resources in Africa." *Water Policy* 1 (1998).

Austin, Mary. *The Land of Little Rain.* New York: Dover, 1903.

Ball, Philip. *Life's Matrix: A Biography of Water.* New York: Farrar, Straus and Giroux, 1999.

Bannister, Anthony, Peter Johnson, and Alf Wannenburgh. *The Bushmen.* Cape Town: Struik, 1979.

Barlow, Maude. *Blue Covenant: The Global Water Crisis and the Coming Battle for the Right to Water.* New York: New Press, 2007.

Barlow, Maude, and Tony Clark. *Blue Gold: The Fight to Stop the Corporate Theft of the World's Water.* New York: New Press, 2002.

Barnard, Alan. *Hunters and Herders of Southern Africa: A Comparative Ethnography of the Khoisan Peoples.* Cambridge: Cambridge University Press, 1992.

Barringer, Felicity. "Lake Mead Could Be Within a Few Years of Going Dry, Study Says." *New York Times,* February 13, 2008.

Bate, Roger. *The Cost of Free Water: The Global Problem of Water Misallocation and the Case of South Africa.* Johannesburg: Free Market Foundation, 1999.

———. "Water: Can Property Rights and Markets Replace Conflict?" Chapter 15 in *Sustainable Development: Promoting Progress or Perpetuating Poverty?* ed. Julian Morris. London: Profile Books, 2002.

Beard, Peter. *The End of the Game*. San Francisco: Chronicle Books, 1965.

Benyus, Janine M. *Biomimicry: Innovation Inspired by Nature*. New York: Perennial, 2002.

Bergkamp, Ger, Brett Orlando, and Ian Burton, eds. *CHANGE: Adaptation of Water Resources Management to Climate Change*. Gland, Switzerland: IUCN, 2003.

Berry, Wendell. *The Unsettling of America: Culture and Agriculture*. San Francisco: Sierra Club Books, 1986.

Biesele, Megan, and Paul Weinberg. *Shaken Roots: The Bushmen of Namibia*. Marshalltown, South Africa: EDA Publications, 1990.

Biesele, Megan. *Women Like Meat: The Folklore and Foraging Ideology of the Kalahari Ju/'hoan*. Bloomington: Indiana University Press, 1993.

Bishop, Kristyna. "Squatters on Their Own Land: San Territoriality in Western Botswana," September 19, 2002, accessed from firstpeoples.org/land_rights/Southern-Africa.

Blakeslee, Nate. "The Secret Life of Sewage." *Texas Monthly*, December 2007.

Bleek, W. H. I., and L. Lloyd. *Specimens of Bushman Folklore*. London: Allen, 1911.

Bond, Patrick. *Unsustainable South Africa: Environment, Development and Social Protest*. Pietermaritzburg: University of Natal Press, 2002.

Borenstein, Seth. "U.S. in for Perilous Weather as World Warms, NASA Says." Associated Press, August 31, 2007.

Bothma, J. du P., ed. *Game Ranch Management*. Pretoria: J. L. Van Schaick, 1989.

Bridgland, Fred. "Forced to March Into Oblivion: As Botswana Insists That the Kalahari Bushmen 'Will Never Be Dispossessed'." *Weekly Telegraph* (London), issue 330.

Brown, Lester R. *Eco-Economy: Building an Economy for the Earth*. New York: Norton, 2001.

———. *Plan B 2.0: Rescuing a Planet Under Stress and a Civilization in Trouble*. New York: Norton, 2006.

Brubaker, Elizabeth. *Property Rights in the Defense of Nature*. London: Earthscan, 1995.

Campbell, Alec C., and J. Coke, eds. *The Management of Botswana's Environment*. Gaborone: Botswana Society, 1984.

Carrels, Peter. *Uphill Against Water: The Great Dakota Water War*. Lincoln: University of Nebraska Press, 1999.

Carruthers, Vincent. *The Wildlife of Southern Africa: A Field Guide to the Animals and Plants of the Region*. Cape Town: Struik, 2000.

Clarke, James. *Coming Back to Earth: South Africa's Changing Environment*. Johannesburg: Jacana, 2002.

Collier, Michael, Robert H. Webb, and John C. Schmidt. *Dams and Rivers: Primer on the Downstream Effects of Dams*. Tucson, AZ: U.S. Geological Survey, 1996.

Cooley, Heather, Peter Gleick, and Gary Wolf. *Desalination, with a Grain of Salt: A California Perspective.* Oakland, CA: Pacific Institute, 2006.

Cook, Edward R., et al. "North American drought: Reconstructions, Causes, and Consequences." *Earth-Science Reviews* 81, 2007.

Corry, Stephen. "Civilization in Reverse." *Mmegi* (Gaborone, Botswana), March 8–14, 2002.

———. "Open Diary of Journey Through the Central Kalahari, March 19–26." Unpublished report, 2004.

Cosgrove, William, and Frank Rijsberman. *Vision Report: Making Water Everyone's Business.* London: Earthscan, 2000.

Coulson, David, and Alex Campbell. *African Rock Art: Paintings and Engravings on Stone.* New York: Harry N. Abrams, 2001.

Curtin, Fiona. "Transboundary Impacts of Dams: Conflict Prevention Strategies." World Commission on Dams, June 30, 2000.

Davies, Bryan, and Jenny Day. *Vanishing Waters.* Cape Town: University of Cape Town Press, 1998.

Depster, Carolyn. "Botswana Bushmen's Last Stand." BBC, March 18, 2002.

De Soto, Hernando. *The Mystery of Capital: Why Capitalism Triumphs in the West and Fails Everywhere Else.* New York: Basic Books, 2000.

De Villiers, Marq. *Water: The Fate of Our Most Precious Resource.* New York: Houghton Mifflin, 2000.

Diamond, Jared. *Collapse: How Societies Choose to Fail or Succeed.* New York: Viking, 2005.

———. *Guns, Germs, and Steel: The Fates of Human Societies.* New York: Norton, 1999.

Dinesen, Isak. *Out of Africa.* New York: Vintage, 1937.

Dubash, N. K., M. Dupar, S. Kothari, and Tundu Lissu. *A Watershed in Global Governance? An Independent Assessment of the World Commission on Dams.* Washington, DC: World Resources Institute, 2001.

Duncan, David James. *My Story as Told by Water.* San Francisco: Sierra Club Books, 2001.

———. *River Teeth: Stories and Writings.* New York: Doubleday, 1995.

Dyson, Megan, Ger Bergkamp and John Scanlon, eds. *FLOW: The Essentials of Environmental Flows.* Gland, Switzerland: IUCN, 2003.

Easterly, William. *The White Man's Burden: Why the West's Efforts to Aid the Rest Have Done So Much Ill and So Little Good.* New York: Penguin, 2006.

Eckholm, Erik. "To Quench China's Thirsty North, Waters and People Will Move." *New York Times,* August 27, 2002.

Economist Intelligence Unit. *Country Report: Botswana.* London: EIU, January 2006.

Emerton, Lucy, and Elroy Bos. *VALUE: Counting Ecosystems as Water Infrastructure.* Gland, Switzerland: IUCN, 2004.

Emoto, Masaru. *The Hidden Messages in Water.* New York: Atria Books, 2001.

Epstein, Edward Jay. *The Diamond Invention.* Also published as *The Rise and Fall of Diamonds: The Shattering of a Brilliant Illusion.* New York: Simon and Schuster, 1982.

Estes, Richard Despard. *The Behavior Guide to African Mammals.* Johannesburg: Russel Friedman Books, 1991.

Fagan, Brian. *The Great Warming: Climate Change and the Rise and Fall of Civilizations.* New York: Bloomsbury Press, 2008.

———. *The Long Summer: How Climate Changed Civilization.* New York: Basic Books, 2004.

Faruqui, Naser I., Askit K. Biswas, and Murad J.Bino, eds. *Water Management in Islam: Water Resources Management and Policy.* New York: United Nations University, 2001.

Fielder, Christine, and Chris King. *Sexual Paradox: Complementarity, Reproductive Conflict and Human Emergence.* Lulu.com, 2006.

Finnegan, William. "Leasing the Rain: The World Is Running Out of Fresh Water, and the Fight to Control It Has Begun." *New Yorker,* April 8, 2002.

Flannery, Tim. *The Weather Makers: The History and Future Impact of Climate Change.* New York: Penguin, 2005.

Forero, Juan. "Latin America Fails to Deliver on Basic Needs." *New York Times,* February 22, 2005.

Freid, Stephanie. "Future of War Will Go with the Flow: Water Promises to Be a Flash Point." *San Francisco Chronicle,* June 10, 2007.

Fuller, Alexandra. *Don't Let's Go to the Dogs Tonight: An African Childhood.* New York: Random House, 2001.

Gall, Sandy. *The Bushmen of Southern Africa: Slaughter of the Innocent.* London: Random House, 2002.

Gasana, James K. "Natural Resource Scarcity and Violence in Rwanda." In *Conserving the Peace: Resources, Livelihoods and Security,* edited by R. Matthew, M. Halle, and J. Switzer. Gland, Switzerland: IUCN, 2002.

Gleick, Peter H. *Dirty Water.* Washington, DC: Island Press, 1996.

———. "Making Every Drop Count." *Scientific American,* February 2001.

———. *The World's Water: The Biennial Report on Fresh Water Resources, 1998–1999.* Washington, DC: Island Press, 1998.

———. *The World's Water: The Biennial Report on Fresh Water Resources, 2002–2003.* Washington, DC: Island Press, 2002.

Gleick, Peter H., Gary Wolff, Elizabeth L. Chalecki, and Rachel Reyes. *The New Economy of Water: The Risks and Benefits of Globalization and Privatization of Fresh Water.* Oakland: Pacific Institute, 2002.

Glennon, Robert. *Water Follies: Groundwater Pumping and the Fate of America's Fresh Waters.* Washington, DC: Island Press, 2002.

Godwin, Peter. "The Bushmen." *National Geographic,* February 2001.

Gordon, Robert J. *The Bushman Myth: The Making of a Namibian Underclass.* Boulder: Westview Press, 1992.

Gourevitch, Philip. *We Wish to Inform You That Tomorrow We Will Be Killed with Our Families: Stories from Rwanda.* New York: Farrar, Straus and Giroux, 1998.

Government of Botswana. *Final Report Management Plan: Central Kalahari Game Reserve.* Gaborone: Department of Wildlife and National Parks, 2001.

Green Cross International. *Water for Peace in the Middle East and Southern Africa.* Geneva: Green Cross International, 2000.

Hall, David. "Water in Public Hands: Public Sector Water Management—A Necessary Option." *Public Services International,* June 2001.

Hamner, Jesse H., and Aaron T. Wolf, "Patterns in International Water Resource Treatises: The Transboundary Freshwater Dispute Database." *Colorado Journal of International Law and Policy,* 1997.

Hanks, John. "The Kalahari Desert." In *Wilderness: Earth's Last Wild Places,* edited by P. Gil. Mexico City: Cemex/Conservation International, 2006.

Harrison, G. A., H. J. M. Tanner, D. R. Pilbeam, and P. T. Baker. *Human Biology.* Oxford, UK: Oxford Science, 1992.

Hayes, Bill. *The Anatomist.* New York: Random House, 2007.

Hessler, Peter. "Underwater: The World's Biggest Dam Floods the Past." *New Yorker,* July 7, 2003.

Heyns, Piet, et al., eds. *Namibia's Water: A Decision Maker's Guide.* Windhoek: Department of Water Affairs, 1998.

Hillel, Daniel. *The Rivers of Eden: The Struggle for Water and the Quest for Peace in the Middle East.* New York: Oxford University Press, 1994.

Hitchcock, Robert K. *Kalahari Communities: Bushmen and the Politics of the Environment in Southern Africa.* Copenhagen: International Work Group for Indigenous Affairs, document no. 79, 1996.

———. "Mobility, Sedentism, and Intensification: Organizational Responses to Environmental and Social Change among the San of Southern Africa." In *Processual Archaeology: Exploring Analytical Strategies, Frames of Reference, and Culture Process,* ed. Amber Johnson. Westport, CT: Praeger, 2004.

———. "Mobility Strategies Among Foragers and Part-Time Foragers in the Eastern and Northeastern Kalahari Desert, Botswana." Unpublished paper, University of Nebraska.

———. "Resources, Rights, and Resettlement among the San of Botswana." *Cultural Survival Quarterly* 22(4).

Hitchcock, Robert K., and Megan Biesele. "Establishing Rights to Land, Language, and Political Representation: The Ju/'hoansi San at the Millennium." In *Tracing the Rainbow: Art and Life in Southern Africa,* edited by Stevan Eisenhofer. Linz, Austria: Arnoldsche Publishers, 2001.

———. "Rights to Land, Language and Political Representation: The Ju/'hoansi San of Namibia and Botswana at the Millennium." Unpublished article.

———. "San, Khwe, Basarwa or Bushmen? Terminology, Identity and Empowerment in Southern Africa." Unpublished article.

Hitchcock, Robert K., and Diana Vinding, eds. *Indigenous Peoples' Rights in Southern Africa.* Copenhagen: International Work Group for Indigenous Affairs, 2004.

Hochschild, Adam. *King Leopold's Ghost.* Boston: Houghton Mifflin, 1998.

Holden, Richard. "Position Paper on Sustainable Sanitation." Appropriate Technology Conference, November 23, 2001.

Homer-Dixon, Thomas F. "Environmental Scarcities and Violent Conflict." *International Security*, summer 1994.

———. *Environmental Scarcity and Global Security.* New York: Foreign Policy Association Headline Series, 1993.

Hunt, Constance Elizabeth. *Thirsty Planet: Strategies for Sustainable Water Management.* New York: Zed Books, 2004.

Integrated Regional Information Networks: In-Depth. *Running Dry: The Humanitarian Impact of the Global Water Crisis.* New York: United Nations, October 2006.

Intergovernmental Panel on Climate Change. *Climate Change, 2001. Contributions of Working Groups I, II, and III to the Third Assessment Report of the Intergovernmental Panel on Climate Change.* New York: Cambridge University Press, 2001.

International Consortium of Investigative Journalists, edited by M. Beelman et al. *The Water Barons: How a Few Private Water Companies Are Privatizing Your Water.* Washington, DC: Public Integrity Books, 2003.

International Rivers. "Spreading the Water Wealth: Making Water Infrastructure Work for the Poor." Annual Dams, Rivers & People report, 2006.

Isaacson, Rupert. *The Healing Land: A Kalahari Journey.* London: Fourth Estate, 2001.

———. "Last Exit from the Kalahari: the Slow Genocide of the Bushmen/San." *OpenDemocracy*, August 28, 2002.

Jacobson, Peter J. Kathryn M. Jacobson, and Mary K. Seely. *Ephemeral Rivers and Their Catchments: Sustaining People and Development in Western Namibia.* Windhoek: Desert Research Foundation of Namibia, 1995.

Jehl, Douglas. *Whose Water Is It?: The Unquenchable Thirst of a Water-Hungry World.* Washington, DC: National Geographic, 2003.

Jobin, William. *Dams and Disease: Ecological Design and Health Impacts of Large Dams and Irrigation Systems.* New York: Routledge, 1999.

Kanfer, Stefan. *The Last Empire: De Beers, Diamonds, and the World.* New York: Farrar, Straus and Giroux, 1993.

Kaplan, Robert D. *Surrender or Starve: Travels in Ethiopia, Sudan, Somalia, and Eritrea.* Revised edition. New York: Vintage, 2003.

———. *The Ends of the Earth: A Journey to the Frontiers of Anarchy.* New York: Vintage, 1996.

Kapuscinski, Ryszard. *The Shadow of the Sun.* New York: Knopf, 2001.

Katz, Richard. *Boiling Energy: Community Healing Among the Kalahari Kung.* Cambridge, MA: Harvard University Press, 1982.

Katz, Richard, Megan Biesele, and Verna St. Denis. *Healing Makes Our Hearts Happy: Spirituality & Cultural Transformation Among the Kalahari Ju/'hoansi.* Rochester, VT: Inner Traditions, 1997.

Kenyon, Paul. "Row over the Bushmen 'Genocide'." *Crossing Continents*, BBC, November 10, 2005.

Khagram, Sanjeev. *Dams and Development: Transnational Struggles for Water and Power.* Ithaca, NY: Cornell University Press, 2004.

Kiema, Kuela. *The Trap to Catch Roy Sesana.* Unpublished manuscript, Gaborone, 2006.

Kiernan, Ben. *Blood and Soil: A World History of Genocide and Extermination from Sparta to Darfur.* New Haven, CT: Yale University Press, 2008.

King, Jackie M., R. E. Tharme, and M. S. de Villiers, eds. *Environmental Flow Assessments for Rivers: Manual for the Building Block Methodology.* Pretoria: Water Research Commission, 1998.

Kinkade-Levario, Heather. *Design for Water: Rainwater Harvesting, Stormwater Catchment, and Alternate Water Reuse.* Vancouver, Canada: New Society, 2007.

Koeppel, Gerald T. *Water for Gotham: A History.* Princeton, NJ: Princeton University Press, 2000.

Krech, Shepard. *The Ecological Indian: Myth and History.* New York: W. W. Norton, 1999.

LaFranier, Sharon. "Sub-Saharan Africa Lags in Water Cleanup." *New York Times,* August 27, 2004.

Lagod, Martin. "We're Running Out of Water." *San Francisco Chronicle,* July 8, 2007.

Lane, Paul, A. Reid, and A. Segobye, eds. *Ditswa Mmung: The Archaeology of Botswana.* Gaborone: Pula Press/Botswana Society, 1998.

Lavelle, Marianne. "Why You Should Worry About Water: How This Diminishing Resource Will Determine the Future of Where and How We Live." *US News & World Report,* June 4, 2007.

Leakey, Richard. *Wildlife Wars: My Battle to Save Kenya's Elephants.* London: Pan Books, 2002.

Lee, Richard B., and Irven DeVore, eds. *Man the Hunter.* Chicago: Aldine, 1968.

Lee, Richard B., Robert K. Hitchcock, and Megan Biesele, eds. *The Kalahari San: Self-Determination in the Desert.* Cultural Survival Quarterly 25(1):8–61, 2002.

Leopold, Aldo. *A Sand County Almanac.* New York: Ballantine, 1986.

Le Roux, Willemien. *Shadow Bird.* Cape Town: Kwela Books, 2000.

Le Roux, Willemien, and Alison White (eds). *Voices of the San.* Cape Town: Kwela Books, 2004.

Leslie, Jacques. *Deep Water: The Epic Struggle over Dams, Displaced People, and the Environment.* New York: Farrar, Straus and Giroux, 2005.

———. "Running Dry: What Happens When the World No Longer Has Enough Freshwater?" *Harper's,* July 2000.

Lewis-Williams, David. *Images of Mystery: Rock Art of the Drakensberg.* Cape Town: Double Storey Books, 2003.

———. *The Mind in the Cave: Consciousness and the Origins of Art.* New York: Thames and Hudson, 2002.

———. "Vision, Power and Dance: The Genesis of a Southern African Rock Art Panel." Lecture, Netherlands Museum of Anthropology and Prehistory, May 8, 1992.

Liebenberg, Louis. *The Art of Tracking: The Origin of Science*. Cape Town: David Philip, 1990.

Lomborg, Bjorn. *The Skeptical Environmentalist: Measuring the Real State of the World*. Cambridge, UK: Cambridge University Press, 2001.

Lovegrove, Barry. *The Living Deserts of Southern Africa*. Cape Town: Fernwood Press, 1993.

MacDonald, Alan, et al. *Developing Groundwater: A Guide for Rural Water Supply*. London: ITDG, 2005.

Main, Michael. *Kalahari: Life's Variety in Dune and Delta*. Johannesburg: Southern Book, 1987.

Mann, Charles. "The Rise of Big Water." *Vanity Fair*, May 2007.

Mapunda, Dr. S. A. "Xoroxo Duxee Report of Post-Mortem Examination to the Station Commander. Botswana Police, Ghanzi. Filed November 11, 2005.

Marshall, Lorna. *The !Kung of Nyae Nyae*. Cambridge, MA: Harvard University Press, 1976.

Martin, Glen. "Water Woes: Projecting 60 years into California's thirsty future." *San Francisco Chronicle Magazine*, January 7, 2007.

Matthiessen, Peter. *The Tree Where Man Was Born*. New York: E. P. Dutton, 1972.

McCully, Patrick. *Silenced Rivers: The Ecology and Politics of Large Dams*. London: Zed Books, 2001.

McGuane, Thomas. "The Heart of the Game." In *The Best of Outside*. New York: Random House, 1997.

Mendelsohn, John, and Selma el Obeid. *Okavango River: Journey of a Lifeline*. Cape Town: Struik, 2004.

———. *Sand and Water: A Profile of the Kavango Region*. Cape Town: Struik, 2003.

Merriweather, Alfred. *Desert Doctor: Medicine and Evangelism in the Kalahari Desert*. Cambridge, UK: Lutterworth Press, 1969.

Midkiff, Ken. *Not a Drop to Drink: America's Water Crisis (And What You Can Do)*. New York: New World Library, 2007.

Millennium Ecosystem Assessment. *Ecosystems and Human Well-Being: Synthesis*. Washington, DC: Island Press, 2005.

Miller, Lee. *Roanoke: Solving the Mystery of the Lost Colony*. New York: Penguin Books, 2000.

Mitchell, Stephen, translation. *The Book of Job*. New York: HarperPerennial, 1987.

Mogae, Festus G. "Statement by His Excellency the President of Botswana Concerning the Relocation of Basarwa from the Central Kgalagadi Game Reserve." Paid advertisement, *Sunday Times* (London), November 24, 2002.

Mogwe, Alice, ed. *Ditshwanelo Focus Seminar Series: Central Kalahari Game Reserve*. Gaborone: Pyramid, March 2002.

Mogwe, Alice. *When Will This Moving Stop?: Report on a Fact-Finding Mission to the Central Kalagadi Game Reserve*. Gaborone: Ditshwanelo, 1996.

———. *Who Was (T)here First? An Assessment of the Human Rights Situation of Basarwa in*

Selected Communities in the Gantsi District, Botswana. Occasional Paper no. 10, Botswana Christian Council. Gaborone: Botswana Christian Council, 1992.

Mogwe, Alice, and Daniel Tevera. "Land Rights of the Basarwa People in Botswana." In *Environmental Security in Southern Africa,* edited by D. Tevera and Sam Moyo. Harare, Zimbabwe: SAPES, 2000.

Molden, David, ed. *Water for Food; Water for Life: A Comprehensive Assessment of Water Management in Agriculture.* London: Earthscan, 2007.

Molden, David, Charlotte De Fraiture, and Frank Rijsberman, "Water Scarcity: The Food Factor." *Issues in Science and Technology,* summer 2007.

Montaigne, Fen. "Water Pressure: The Earth's Six Billion People Already Overtax Its Supply of Accessible Fresh Water. What Happens When the Planet Gets a Few Billion More Hands?" *National Geographic,* August 2003.

Motseta, Sello. "Government of Botswana Restricts Movement in Kalahari Game Reserve." Associated Press, September 2, 2005.

Mphinyane, Sethunya Tshepo. "Power and Powerlessness: When Support Becomes Overbearing—the Case of Outsider Activism in the Resettlement Issue of the Basarwa of the Central Kalahari Game Reserve." *Pula: Botswana Journal of African Studies* 16, no. 2 (2002).

Nabhan, Gary Paul. *The Desert Smells Like Rain.* San Francisco: North Point Press, 1982.

NASA, Goddard Institute for Space Studies Surface Temperature Analysis, data.giss.nasa.gov/gistemp.

Nash, Roderick. *Wilderness and the American Mind.* Revised edition. New Haven, CT: Yale University Press, 1973.

Newman, Kenneth. *Birds of Southern Africa.* Cape Town: Struik, 2000.

O'Connell, Caitlin. *The Elephant's Secret Sense.* New York: Free Press, 2007.

O'Driscoll, Patrick. "A Drought for the Ages." *USA Today,* June 8, 2007.

Ohlsson, Leif, ed. *Hydropolitics: Conflicts over Water as a Development Constraint.* London: Zed Books, 1995.

Omer-Cooper, J. D. *History of Southern Africa.* London: James Currey, 1986.

Ortega y Gasset, Jose. *Meditations on Hunting.* Translated by Howard B. Westcott. New York: Scribner, 1972.

Outwater, Alice. *Water: A Natural History.* New York: Basic Books, 1996.

Owens, Mark, and Delia Owens. *Cry of the Kalahari.* New York: Houghton Mifflin, 1993.

Pakenham, Thomas. *The Scramble for Africa: The White Man's Conquest of the Dark Continent from 1876 to 1912.* New York: Random House, 1991.

Pallet, John. *Sharing Water in Southern Africa.* Windhoek: Desert Research Foundation of Namibia, 1997.

Pearce, Fred. *The Dammed: Rivers, Dams and the Coming World Water Crisis.* London: Bodley Head, 1992.

———. *Keepers of the Spring: Reclaiming Our Water in an Age of Globalization.* Washington, DC: Island Press, 2004.

———. *When the Rivers Run Dry: Water—the Defining Crisis of the Twenty-First Century.* Boston: Beacon Press, 2006.

Peet, John. "Priceless: A Survey of Water." *Economist,* July 19, 2003.

Phillips, Barnaby. "Bitter Dispute over Bushmen Lands." BBC News, November 24, 2005.

Pollan, Michael. *In Defense of Food: An Eater's Manifesto.* New York: Penguin, 2008.

———. *The Omnivore's Dilemma: A Natural History in Four Meals.* New York: Penguin, 2007.

Postel, Sandra. *Last Oasis: Facing Water Scarcity.* New York: W. W. Norton, 1997.

———. *Pillar of Sand: Can the Irrigation Miracle Last?* New York: W. W. Norton, 1999.

Postel, Sandra, and Brian Richter. *Rivers for Life: Managing Water for People and Nature.* Washington, DC: Island Press, 2003.

Potts, Malcolm, and Thomas Hayden. *Sex and War: How Biology Explains Warfare and Terrorism and Offers a Path to a Safer World.* New York: Benbella Books, 2008.

Power, Samantha. *A Problem from Hell: America and the Age of Genocide.* New York: Basic Books, 2002.

Quammen, David. *Song of the Dodo: Island Biogeography in an Age of Extinction.* New York: Scribner, 1997.

Reader, John. *Africa: A Biography of the Continent.* New York: Random House, 1997.

Reisner, Marc. *Cadillac Desert: The American West and Its Disappearing Water.* Revised and updated. New York: Penguin, 1993.

Ridgeway, Rick. *The Shadow of Kilimanjaro: On Foot Across East Africa.* New York: Henry Holt, 1998.

Ridley, Matt. *The Origins of Virtue: Human Instincts and the Evolution of Cooperation.* London: Penguin Books, 1996.

———. *The Red Queen: Sex and the Evolution of Human Nature.* New York: Macmillan, 1994.

Rifkin, Jeremy. *Beyond Beef: The Rise and Fall of the Cattle Culture.* New York: Plume, 1993.

Roach, Mary. *Stiff: The Curious Lives of Human Cadavers.* New York: Norton, 2003.

Roberts, Janine. *Glitter & Greed: The Secret World of the Diamond Cartel.* New York: Disinformation, 2003.

———. "Masters of Illusion." *Ecologist,* January 9, 2003.

Roodt, Veronica. *Common Wildflowers of the Okavango Delta: Medicinal Uses and Nutritional Value.* Gaborone: Shell, 1998.

———. *Trees & Shrubs of the Okavango Delta: Medicinal Uses and Nutritional Value.* Gaborone: Shell, 1998.

Ross, Karen. *Okavango: Jewel of the Kalahari.* Cape Town: Struik, 2003.

Rothfelder, Jeffrey. *Every Drop for Sale: Our Desperate Battle over Water in a World About to Run Out.* New York: Penguin Putnam, 2001.

Royte, Elizabeth. *Bottlemania: How Water Went on Sale and Why We Bought It.* New York: Bloomsbury, 2008.

Rush, Norman. *Mating.* New York: Vintage, 2001.

Sadoff, Claudia, Thomas Greiber, Mark Smith, and Ger Bergkamp, eds. *SHARE: Managing Water Across Boundaries*. Gland, Switzerland: IUCN, 2008.

Sahlins, Marshall. *Stone Age Economics*. London: Tavistock Publications, 1972.

Saleth, R. Maria, and Ariel Dinar. *The Institutional Economics of Water: A Cross-Country Analysis of Institutions and Performance*. Washington, DC: IBRD/World Bank, 2004.

Saltzman, James. *Thirst: A Short History of Drinking Water*. Raleigh, NC: Duke Law School Faculty Scholarship Series, 2006.

Saugestad, Sidsel. "Improving Their Lives: State Policies and San Resistance in Botswana." *Before Farming*, April 2005.

———. *The Inconvenient Indigenous: Remote Area Development in Botswana, Donor Assistance, and the First People of the Kalahari*. Uppsala, Sweden: Nordic Africa Institute, 2001.

Schoeman, P. J. *Hunters of the Desert Land*. Cape Town: Howard Timmins, 1957.

Scudder, Thayer. *The Future of Large Dams: Dealing with Social, Environmental and Political Costs*. London: Earthscan, 2005.

Scully, Malcolm G. "The Politics of Running Out of Water." *Chronicle of Higher Education*, November 17, 2000.

Sesana, Roy. "What Kind of Development Is This?" Address given before receiving the Right Livelihood Award, Stockholm, December 9, 2005.

Sheller, Paul. *The People of the Kalahari Game Reserve*. Unpublished Report for the Ministry of Local Government Lands and Housing. Gaborone, 1976.

Shiva, Vandana. *Water Wars: Privatization, Pollution, and Profit*. Cambridge: South End Press, 2002.

Shostak, Marjorie. *Nisa: The Life and Words of a !Kung Woman*. Cambridge, MA: Harvard University Press, 1981.

Silberbauer, George S. *Hunter and Habitat in the Central Kalahari Desert*. Cambridge, UK: Cambridge University Press, 1981.

———. *The Bushmen Survey Report to the Bechuanaland Government*. Gaborone: Government Press, 1965.

Simon, Paul. *Tapped Out: The Coming World Crisis in Water and What We Can Do About It*. New York: Welcome Rain, 1998.

SIWI, IUCN, IWMI, IFPRI. *Let It Reign: The New Water Paradigm for Global Food Security*. Working Report to the Commission on Sustainable Development-13, Stockholm International Water Institute, 2004.

Skaife, S. H. *African Insect Life*. Revised and illustrated edition. Cape Town: Struik, 1979.

Skelton, Paul. *A Complete Guide to Freshwater Fishes of Southern Africa*. Cape Town: Struik, 2001.

Smith, Adam. *An Inquiry into the Nature and Causes of the Wealth of Nations*. Chicago: University of Chicago Press, 1976.

Smith, Mark, Dolf de Groot, Daniele Perrot-Maitre, and Ger Bergkamp, eds. *PAY: Establishing Payments for Watershed Services*. Gland, Switzerland: IUCN, 2006.

Snitow, Alan, Deborah Kaufman, and Michael Fox. *Thirst: Fighting the Corporate Theft of Our Water*. New York: John Wiley & Sons, 2007.

Specter, Michael. "The Last Drop: Confronting the Possibility of a Global Catastrophe." *New Yorker*, October 23, 2006.

Speth, James Gustave. *The Bridge at the Edge of the World: Capitalism, the Environment, and Crossing from Crisis to Sustainability*. New Haven, CT: Yale University Press, 2008.

———. *Red Sky at Morning: America and the Crisis of the Global Environment*. New Haven, CT: Yale University Press, 2005.

Stahle, David W., et al. "The Lost Colony and Jamestown Droughts." *Science* 280, 1998.

Stark, Peter. *Last Breath: The Limits of Adventure*. New York: Ballantine Books, 2001.

Stegner, Wallace. *Beyond the 100th Meridian*. New York: Penguin, 1992.

Steinhauer, Jennifer. "Water-Starved California Slows Development; Law Requires 20-Year Supply." *New York Times*, June 7, 2008.

Stevenson, Mark. "Government Responsible for Good Water, Forum Says." Associated Press, March 24, 2006.

Stockholm International Water Institute. "Making Water a Part of Economic Development: The Economic Benefits of Improved Water Management and Services." Stockholm: SIWI, 2005.

Sun Tzu. *The Art of War*. Translated from the Chinese by Thomas Cleary. Boston: Shambhala, 1988.

Survival International. "Bushmen Aren't Forever: Botswana: Diamonds in the Central Kalahari Game Reserve and the Eviction of Bushmen." London, April 7, 2003.

———. "Submission to UN Human Rights Council on Indigenous Peoples and Access to Water." London, April 11, 2007.

Suzman, James. "A Future Beyond Survival? The Interferences of a European Lobby Group Has Worsened the Plight of the Central Kalahari San." *Mail & Guardian* (South Africa), December 13, 2002.

———. "Indigenous Wrongs and Human Rights: National Policy, International Resolutions and the Status of the San of Southern Africa." Windhoek, Namibia: Legal Assistance Center, 2001.

———. "Kalahari Conundrums: Relocation, Resistance and International Support in the Central Kalahari Botswana." *Before Farming* 4 (2002–03).

Tagliabue, John. "As Multinationals Run the Taps, Anger Rises over Water for Profit." *New York Times*, August 26, 2002.

Taylor, I., and G. Mokhawa. "Not Forever: Botswana, Conflict Diamonds and the Bushmen." *African Affairs*, April 2003.

Theroux, Paul. *Dark Star Safari: Overland from Cairo to Cape Town*. London: Penguin, 2002.

Thomas, Elizabeth Marshall. *The Harmless People*. New York: Knopf, 1959.

———. *The Old Way: A Story of the First People*. New York: Farrar, Straus and Giroux, 2006.

Thoreau, Henry David. *Walden* and *Civil Disobedience*. New York: Signet Classics, 1980.

Tlou, Thomas, and Alec Campbell. *History of Botswana*. Gaborone: Macmillan, 1997.

Townsend, Jeff. "Botswana: The San (Bushmen) Rights Case." *Congressional Research Service Report for Congress*. Washington, DC: Library of Congress, October 19, 2004.

Turner, Jack. *The Abstract Wild*. Tucson: University of Arizona Press, 1996.

Turton, Anthony, Peter Ashton, and Eugene Cloete, eds. *Transboundary Rivers, Sovereignty and Development: Hydropolitical Drivers in the Okavango River Basin*. Geneva: Green Cross International, 2003.

Turton, Anthony, and Roland Henwood. *Hydropolitics in the Developing World: A Southern African Perspective*. Pretoria: African Water Issues Research Unit, 2002.

Turton, Anthony. "Water and Conflict in an African Context." *Conflict Trends* 5 (1999).

Tutu, Desmond. "Statement in Support of the Bushmen of Botswana." Video. Accessed at www.iwant2gohome.org.

UNDP. *Beyond Scarcity: Power, Poverty and the Global Water Crisis*. Summary of the Human Development Report, 2006.

USDA. *Agricultural Resources and Environmental Indicators* 2000. Washington, DC, 2000.

Van der Post, Laurens. *First Catch Your Eland: A Taste of Africa*. London: Hogarth Press, 1977.

———. *Heart of the Hunter*. New York: Viking, 1961.

———. *The Lost World of the Kalahari*. New York: Viking, 1958.

———. *A Mantis Carol*. Washington: Island Press, 1975.

Van Rooyen, Noel. *Flowering Plants of the Kalahari Dunes*. Cape Town: Ekotrust, 2001.

Veth, Peter, Mike Smith, and Peter Hiscock, eds. *Desert Peoples: Archaeological Perspectives*. Oxford, UK: Blackwell, 2005.

Vidal, John. "Australia suffers worst drought in 1000 years: Depleted reservoirs, failed crops and arid farmland spark global warming tussle." *Guardian* (London), November 8, 2006.

Vision 21. *Water for People: A Shared Vision for Hygiene, Sanitation and Water Supply*. Geneva: Water Supply and Sanitation Collaborative Council, 2000.

Ward, Diane Raines. *Water Wars: Drought, Floods, Folly and the Politics of Thirst*. New York: Riverhead Books, 2002.

Watson, Lyall. *Elephantoms: Tracking the Elephant*. New York: Norton, 2002.

Weaver, Tony. "Going Back to their Roots." *Mail & Guardian* (South Africa), August 31, 2001.

Weinberg, Paul. *Once We Were Hunters: A Journey with Africa's Indigenous People*. Cape Town: David Philip, 2000.

Wells, Spencer. *The Journey of Man: A Genetic Odyssey*. Princeton, NJ: Princeton University Press, 2003.

White, John. "Return to Ronoake" (1590). In Richard Hakluyt, *Principal Navigations, Voyages of the English Nation, vol. III*. London, 1600.

Wiessner, Polly. "Banking Time and Banking Relationships: Perspectives from Anthropology, Ethology, and Neuroscience." Presentation at Time Banking Congress, Toronto, August 2004.

———. "Hunting, Healing and Hxaro Exchange: A long-term perspective on !Kung Large Game Hunting." *Evolution of Human Behavior* 23 (2002).

———. "!Kung San Networks in a Generational Perspective." In *The Past and Future of !Kung Ethnography: Critical Reflections and Symbolic Perspectives Essays in Honor of Lorna Marshall*. Edited by Megan Biesele with Robert Gordon and Richard Lee. Hamburg: Helmut Buske Verlag, 1986.

———. "Measuring the Impact of Social Ties on Nutritional Status among the !Kung San." *Social Science Information* 20 (1981).

———. "Owners of the Future? Calories, Cash, Casualties and Self-Sufficiency in the Nyae Nyae Area Between 1996 and 2003." *Visual Anthropology Review* 19, Issue 1–2 (2004).

———. "Risk, Reciprocity and Social Influences on !Kung San Economics." In *Politics and History in Band Societies*. Edited by Eleanor Leacock and Richard B. Lee. Cambridge: Cambridge University Press, 1982.

Williams, Terry Tempest. *Refuge: An Unnatural History of People and Place*. New York: Pantheon Books, 1991.

Wilmsen, Edwin. *Land Filled with Flies: A Political Economy of the Kalahari*. Chicago: University of Chicago Press, 1989.

Wilson, Bee. "The Last Bite: Is the World's Food System Collapsing?" *New Yorker*, May 19, 2008.

Wilson, David Sloan. *Darwin's Cathedral: Evolution, Religion, and the Nature of Society*. Chicago: University of Chicago Press, 2002.

Wolf, Aaron T. "Conflict and Cooperation along International Waterways." *Water Policy* 1, no. 2 (1998).

———. "Development and Transboundary Waters: Obstacles and Opportunities." Submitted to the World Commission on Dams, July 2, 2000.

Wolff, G., and E. Hallstein. *Beyond Privatization: Restructuring Water Systems to Improve Performance*. Oakland, CA: Pacific Institute, 2005.

Wood, Chris. *Dry Spring: The Coming Water Crisis of North America*. Vancouver, BC: Raincoast Books, 2008.

Working Group on Indigenous Minorities in Southern Africa. *The Khwe of the Okavango Panhandle: The Past Life*. Ghanzi, Botswana: Teemacane Trust, 2002.

Workman, James G. "How to Fix Our Dam Problems." *Issues in Science and Technology*, Fall 2007.

World Bank. *Assessment Report: Complaint Regarding IFC's investment in Kalahari Diamonds Ltd, Botswana*. Washington, DC: Office of the Compliance Advisor/Ombudsman International Finance Corporation/Multilateral Investment Guarantee Agency, June 2005.

World Commission on Dams. *Dams and Development: A New Framework for Decision-Making*. London: Earthscan, 2000.

Worster, Donald. *Rivers of Empire: Water, Aridity, and the Growth of the American West.* New York: Oxford University Press, 1985.

Wright, Robert. *The Moral Animal: Why We Are the Way We Are; Evolutionary Psychology and Everyday Life.* New York: Pantheon, 1994.

———. *Nonzero: The Logic of Human Destiny.* New York: Vintage, 2001.

Xu, Yongxin, and Hans E. Beekman, eds. *Groundwater Recharge Estimation in Southern Africa.* Paris: UNESCO, 2003.

Yadin, Yigael. *Masada: Herod's Fortress and the Zealots' Last Stand.* New York: Random House, 1966.

Zavis, Alexandra. "Driven to Near Extinction: An Ancient Tribe Is Battling to Remain in a Wildlife Reserve." Associated Press, January 10, 2006.

Zito, Kelly. "Dry Spell Danger: Skimpy Snowpack, Low Rainfall, Growing Populations and Demands of Agriculture and Recreation Add Up to an Urgent Need for Water Conservation as Californians Stare into the Face of Drought." *San Francisco Chronicle,* May 2, 2008.

Acknowledgments

The burden of the ignorant and landless squatter is a growing problem in Africa—and one I imposed on others wherever I went. Nomadic years spent living out of a Land Rover are as carefree as they are lonely and disorienting, and while utter freedom generates wild experiences it can also generate nervous breakdowns. Only in hindsight does one see how near the edge I wandered, so my first gratitude goes to those who anchored me with a home-cooked meal, bracing conversations, or a traveling companion when I arrived on their threshold without warning. For such quintessentially African hospitality I am especially grateful to the families of Hamilton Wende, Paul Weinberg, Chaminda Rajapakse, Bruce Lawson, Eddie Koch, Paul Walters, Nick Charalambides, Alex Hetherington, Paul Robertson, Lindsay Bolus, Liz Rihoy, Pamela Paul, Christopher Vaughn, Verona Smith, and Sophie Simmonds. When my travels landed me back briefly in the United States, I found that Becky King, George Schwimmer, Peter Mali, Eric Kessler, Hedrick Belin, Alex Baldwin, and Greg Widmyer still made welcome room despite new marriages and children as their former roommate imposed himself yet again. Naturally, those with the least gave me the most. The families of Qoroxloo, Mongwegi, Gakelekgolele, Moagi, Gaoberekwe, Sesana, Amogelang, and Nyare Bapalo were constantly on the verge of losing their home in the Kalahari, and further increased their risks of retaliation by hosting me, yet did not hesitate to invite me to their hearth to share their game meat, tsama melons, and stories. All these individuals helped transform a safari, which is a mere trip in the wild, into a journey of discovery enriched by the warmth of people.

One could not ask for better mentors at the start of my explorations into the turbulent politics of water scarcity than Bruce Babbitt, John Leshy, and David Hayes, whose years of experience tempered my passion without in any way suppressing it. Upon my arrival in Africa, my colleagues at the World Commission on Dams helped me appreciate the global extent of the social and economic risks inherent to the rule of water. In seeking consensus, chairman Kader Asmal, in addition to introducing me to my future wife, pushed the edge of redefining the art of the possible. Commissioners Deborah Moore, Ted Scudder, Don Blackmore, and Medha Patkar often sat down with me to share their sense of water-borne tensions within and between nations. In the no-man's-land of highly politicized river and dam debates, the unflappable Achim Steiner led our battle-scarred band of colleagues—Bruce Aylward, Jeremy Bird, Chris Clarke, Kate Dunn, Larry Haas, Mark Halle, Sanjeev Khagram, Madiodio Niasse, S. Parasuraman, Corli Pretorius, John Scanlon, Jamie Skinner, and the late Saneeya Hussain—into the fray, forging a bond that endures to this day.

Mark Twain once urged writers: "Get your facts first. Then you may distort them as much as you please." As a work of narrative nonfiction *Heart of Dryness* is not scholarly, but it draws from the deep wellspring of peer-reviewed publications that are. In the two often unrelated fields of hydrology and anthropology, acomplished individuals went beyond their busy schedules to take me under their wing and provide a realistic and scientific grounding for my energy, questions, confusion, and enthusiasms.

In the world of water, I'm grateful to those who encouraged my investigations and enriched it with their own research. Anthony Turton linked state sovereignty to control over its waters while Aaron Wolf and Peter Gleick illustrated how, why, and where a water-stressed nation will not war over water with another sovereign state, but like water may well attack and break down that which is weaker than itself. Peter Ashton illustrated the transcending ties between people, ecology, and fluctuating currents, and Bryan Davies and Jackie King enhanced my appreciation for how intricately, thoroughly, and dangerously our species rests on the health of rivers. Mary Seely showed me the myriad ways desert life, human or otherwise, relies on the capture of fleeting metabolic drops of water. Chris Schaan walked me through his documentation of where, when, how, and why a nation squandered its water. Sharon Pollard and Guy Preston explained how empowering all people with ownership of water can benefit not only the poor but the natural foundation shared by all. Anton Earle brought the concept of "virtual water" down to a level where it could be tasted. Christine Colvin unearthed the complex dynamics of unseen water locked beneath the surface. Ger Bergkamp and Mark Smith became vigorous sparring partners in robust debates over the emerging political implications of water stress and scarcity. Greg Thomas, Brianna Randall, Laura Wildman, and Monty Simus opened the menu of ways to transform existing water infrastructure.

I was equally lucky in the field of Bushmen studies. If indeed 80 percent of success is just showing up, years wandering in Africa let me cross paths with some of the most successful authorities on Bushmen politics, anthropology, and archaeology. I was fortunate to meet—and be set straight by—the irascible and myth-busting John Marshall in Namibia, shortly before his death. Jamie Traut, Louis Liebenberg, Bruce Lawson, Smith Moeti, and Galomphete Gakelekgolele each taught me, in different ways, to track, smell, listen, and move in the Kalahari, thinking like a hunter. In Gaborone Michael Taylor gave me the opportunity to work closely on another Bushmen-related book project with the incomparable Alec Campbell, whose experience and quiet command of material channeled my neophyte fascination with Botswana's past. Alec introduced me to Michael Main, Larry Robbins, George Brook, Jim Denbow, Nick Walker, Phillip Segadika, and Ed Wilmsen, whose understanding of the habitat and heritage of Bushmen and of the Kalahari enriched my own sense of people and of place. Didier Bouakaze-Khan brought me face-to-face with the creative legacy of Bushmen at Tsodilo Hills, then showed how David Lewis-Williams could interpret rock art from the other side. From Botswana links I also came to know the definitive work and lively character of George Silberbauer, whose writing lay the foundation for my own efforts. Finally the perceptive field researchers and eminent scholars Polly Wiessner, Megan Biesele, and Robert Hitchcock each provided moral support for this book and offered essential insights into how Bushmen use, define, recognize, and exchange rights related to property and water resources. Dr. S. A. Mapunda connected the consequences of the loss of water by Bushmen unfolding in the tissues of one single extraordinary woman. I'm also grateful for the time and energy extended to me by those who have been more directly engaged in advocacy work on behalf of defining the place and the rights of today's Bushmen in Botswana, including Paul Sheller, Arthur Albertson, Alice Mogwe, Glyn Williams, Kali Mercier, Fiona Watson, Miriam Ross, Gordon Bennett, Kuela Kiema, and

Maitseo Bolaane. All these people may at times have disagreed vehemently with each other—and in particular with me—but through our interactions and interviews they each augmented my grasp of the past and present status, dilemmas and aspirations of individual and collective groups of Bushmen.

There is not always a clear moment when a dormant seed becomes a living organism, but several uniquely American institutions provided the resources to nurture nascent thoughts into the conclusions found in this book. I could not even have first conceived this idea without the financial and institutional support of the Institute of Current World Affairs; in particular Peter and Lu Martin, who raised donor funds that gave me a precious period of time to conduct primary research in the field, in the right place at the right time. A fellowship at the Kinship Conservation Institute, thanks especially to Karie Thomson, Brent Heglund, and Jack Loacker, revealed water and nature in an economic light, while the Property and Environment Research Center's Bobby McCormick and Terry Anderson helped me see human nature, warts and all, as potential force for change if harnessed through a human right to water.

I'm also indebted to the encouragement of several gifted and humane writers: Robert Wright, David James Duncan, David Quammen, and Ben Skinner. Ben led me to the door of his and now my own infinitely patient yet persistent agent/therapist Geri Thoma, who in turn brought me to the attention of the indefatigable George Gibson, who as editor shaped the book's structure, style, and content from the start, and who along with Michele Lee Amundsen tactically and tactfully helped steer me back toward the destinations I had unknowingly been seeking all along. Also at Walker, I'm grateful for the intestinal fortitude of Vicki Haire and Greg Villepique, who patiently and painstakingly rescued me from countless embarrassing mistakes I had no idea I was making but from which I have hopefully learned.

My oldest gratitude goes out to Gil and Nancy Workman, who were woken up at 3:47 one morning to hear a strained, scratchy voice calling long distance, coming through in short patchy bursts—"hey . . . it's your son . . . listen, I can't talk long cause this cell phone may die . . . but I just want to say I'm hanging on . . . still alive, and while the Kalahari has been sealed off I might be able to"—before the line went dead. It was another eighteen hours before I could reach them again, and I am only now as a new parent beginning to see what such silent hours can do to the psyche. To my wife Vanessa, *le maire de mon coeur, la femme qui boit à l'eau,* my first editorial critic and mother of the two suns who eclipse all other creative endeavors, thank you for sharing the journey, and for asking directions as the stubborn driver gets lost along the way.

Many of the people gratefully acknowledged above may dispute or disagree with the book's premises, analyses, or conclusions. Such is the nature of what will likely become an even more contentious debate as drought worsens, competition grows, and tensions escalate in the politics of running out of water. That said, let there be no guilt by association. In the spirit of Mark Twain's advice let me add that responsibility for the original facts are foremost theirs; all subsequent errors and distortions belong entirely to the author.

Index

A Note on the Author

James Workman, a graduate of Yale and Oxford, began his award-winning career as a journalist in Washington, D.C., writing for the *New Republic, Washington Monthly, Utne Reader, Washington Business Journal, Foreign Service,* and *Orion,* among other publications. In the Clinton administration he served as a special assistant and natural resources speechwriter to Interior Secretary Bruce Babbitt, with whom he reintroduced wolves, restored fire to its vital and natural role in western forests, and blew up obsolete dams to replenish dying rivers. For seven years in Africa and Asia he helped prepare and launch the landmark *Report of the World Commission on Dams,* filed monthly dispatches on water scarcity, led investigative research safaris, lectured at universities, and advised businesses, aid agencies, and conservation organizations on water policy in the developing world. Based on his experience with the Kalahari Bushmen, he is pioneering new platforms for trading the human right to water. He lives with his wife, Vanessa, and two daughters, Camille and Louise, in San Francisco, where he is at work on his second book.